PLANET EARTH DEMANDS

ENERGY, ECONOMICS, EMPLOYMENT, AND OUR INNER AND OUTER ENVIRONMENTS

SHAMCHER BRYN BEORSE

ℭℬ

THE SHAMCHER ARCHIVES

ALPHA GLYPH PUBLICATIONS

Planet Earth Demands
© 1980 by Bryn Beorse and © 2014 The Shamcher Archives
Introduction © 2014 by Carol Sill
Cover Design: Diane Feught

Library and Archives Canada Cataloguing in Publication

Beorse, Bryn, 1896-1980, author
    Planet earth demands : energy, economics, employment and our inner and outer environments / Shamcher Bryn Beorse.

ISBN 978-0-9781705-9-2 (pbk.)

    1. Ecology. 2. Human ecology. I. Title.

QH541.B46 2014   577.27   C2014-902002-3

Alpha Glyph Publications Ltd.   Vancouver BC Canada
www.alphaglyph.com
www.shamcher.org, www.planet-earth.shamcher.com

"Planet Earth, itself, now demands what the sages and prophets shouted in vain."

Contents

INTRODUCTION

Economist, engineer, generalist and mystic, Shamcher Bryn Beorse, reveals his comprehensive overview of the forces and influences shaping humanity in the latter half of the 20th century. Concerned with the fate of the earth environmentally, socially and politically, he offered both advice and warning, peppered with personal anecdotes. The cry of mother earth, the complexity of social issues, and the needs and desires of human beings living in this world today all combine in Beorse's bird's eye view.

This expanded and all-inclusive vision of the cry of the earth is as important today as when it was written over 40 years ago. What seemed radical at that time is commonplace today - an awareness of the totality of the environment including ourselves as well as the development of the inner life. In Beorse's world-view there is no separation between the areas of energy, economics, employment, the individual's pursuit of happiness and his own personal life-experiences. He subtly includes the spiritual life, touching on yoga and Sufi thought and practice as necessary and meaningful tools to address our current problems - not only at a personal level, but in the areas of city life, the environment, education and the media.

He puts his message into simple everyday language of the time, and makes sense of daily life in the broader environmental context. But this book is not merely an overview or a philosophical explanation. It is an urgent call for help from Earth Herself. As a serious meditator, Beorse had heard this cry throughout his life and dedicated his expertise and abilities to answering it. He travelled the world as an engineer and economist, finally settling in the US in the 1950s to work on OTEC, Ocean Thermal Energy Conversion, a system that produces benign solar power from the sea. At the time of writing this book, OTEC had yet to be adopted as an energy system of choice; instead coal, oil, and nuclear interests held the center stage despite OTEC's proof, promise and possibility. Beorse emphasizes this technology as one of the answers to the demand of the planet.

Inside the book, each chapter connects with a specific aspect of the totality of our environment. At first poetic, the first chapter, *Who is Mother Earth?*, is seen from a planetary view, but zooms in to an individual at the corner of Wall St. and Broadway. Soon after it shifts to introducing OTEC, which is immediately followed by the economic chapter, *Money, Money, Money,* and includes many excerpts from the writings of others in the field. *The Subtle Game of Choice* again picks up on OTEC as an energy system, also discussing politics. The chapter on education, *Are You Educated?*, examines experience from birth to school to the world at large, finishing with a spiritual teaching tale. Also ending with a Sufi story, the chapter on our dense social environment, *What is America All About?*, looks at our work environment and touches on the tragedy of involuntary unemployment. *Making Cities Liveable* discusses people and urban planning, while *Lover and Beloved* asks the question: "How can we hope to manage our common environment if we cannot even stand each other's idiosyncrasies?" Beorse touches on the rich variety of human experience and the love of these individual qualities, ending with a tale of his encounter with the King of Afghanistan. In *The Swallowed Environment*, discussion of diet soon gives way to yogis, mystics and non-eating saints.

Our overwhelming information and entertainment environment is the main topic of *The Printed, the Videoed and Audioed,* while accounts of personal experiences in Australia and Borneo in *Neighbor* reveal personal economics, sharing and understanding. *A Closer Neighbor: Your Body,* goes into acupuncture, yoga and Sufi approaches to knowledge and awareness of the body. *The Gold Mine Between Our Ears* is the mind, here seen as the source and substance of our non-physical environment. Yoga and Sufism are discussed alongside mind training and brainstorming. The next chapter, *Beyond Mind?*, examines out of body experience and transfer of consciousness, through his true story of an adventure in Dayakland.

Inter-environment shows how one phase of the environment influences and merges with other phases, and through true wartime stories reveals non-hierarchical working as an effective way of channelling this interpenetration of environmental phases. *Are You Emotional?* discusses emotions, IQ, intuition and telepathy, and mentions an encounter with the Sufi teacher, Inayat Khan as a vibrant example of thought-transfer. Questioning "loyalty" and tackling the folly of top-down organization, *This Delightful Disobedience* shows the shift from theory to insight. Celebrating generalists and their influence, *Comprehensive Designer* examines the work of Buckminster Fuller, the benefits of small ventures and the power of the individual computer.

The final chapter, *Bird's Eye View,* is the overview that shows through study and story the need for comprehensive ecological awareness and action. It ends the book with a sad tale of the effect of involuntary unemployment, revealing how one man's experience has left an impression in nature and the environment.

As an elder and Sufi, Beorse taught and guided many younger people of the day to find the inner guidance that would help them develop their life's purpose. He was insistent that the time for one-sided passive meditation was no more, and that the inner spirit of guidance, our intuition, was to be awakened and developed to serve humanity. Just as technology is the application of science, so is serving humanity the natural application of the meditative life. All we have to do is learn to listen, and put intuition into action.

Beorse points us to the yogis he describes in the book who do not identify themselves with their bodies but see the whole environment, the whole planet, as themselves.

The personal anecdotes that he shares here can be seen in the same light. Not only illustrations of wider principles, the autobiographical accounts are the book's backbone, revealing an approach that radically defies all topic categorization and specialization.

It has been nearly 40 years since Shamcher wrote this book which he revised and kept current until his passing in 1980. Are we ready now to respond?

It was his love of this life and of this precarious human experiment that urged him to write and add his voice to the increasingly urgent call of our planet. At the center of this book is a message of a love that dissolves all the seeming separations between the various disciplines he addresses here. With love as the integrating force and intuition as our inner guidance, humankind will be able to answer the demands of Planet Earth, and survive.

Carol Sill, Vancouver BC, 2014

WHO IS PLANET EARTH?

D o I share with you my relationship with that remarkable lady? Are you one of her fans also? Of considerable age and wisdom, she counsels all who want to listen. She has talked to me since I was a child. With her, I am still a child. I see no authority, no source of knowledge comparable to hers. Yet, she sits in no palace, not even in a White House. "I don't need a palace to entertain beggars of favours." Her father, the Sun, the whole solar system, the galaxies, the Universe, visible and invisible, all, all are her family, source of inspiration which she bountifully shares.

In her bowels and on her surface energy bubbles. She became a bit tired of the over-use and misuse of oil and nucelonics so she sent an engineer (myself) to France to study the ocean energies. Back in the U.S. she guided me gently to other friends of hers – Dr. James Hoffman of the National Bureau of Standards, Professor Everett D. Howe of the University of California. When the nation failed to catch on and ignored our findings and suggestions, Planet Earth bided her time and then dug up more friends: Robert Cohen of the National Science Foundation; William Heronemus of the University of Massachusetts; Lloyd Trimble and Bernie Messinger of the Lockheed Missile and Space Company. In Houston, in May, 1975. five hundred watched one hundred thirty of us hustle and put the pieces together.

Parallel with this development she had guided me to economists who knew how to handle the impact of new energy sources and who saw the flow of goodies, services, money and credit cards be- yond and above theories and forecasts. Forecasts strangle us, stymie action, displease Planet Earth. Our future is made through action, not forecasts.

My engineering and economist friends, my wealth of experience – all are appreciated though not even comparable to the impact of Planet Earth's direct guidance. No wonder our Indian brothers have called her Mother Earth, and my late physicist-philosopher friend, Oliver Reiser, saw her as a giant living creature in which we all lived, moved, worked, burped and made merry. Ben Franklin knew her and so did Abe Lincoln. My most treasured professional and non-professional friends of this day know her.

Planet Earth and her ancestry are concerned not merely with energy, food and adequate ecology, which are just means to an end; means to give us time and strength to delve into the mystery and joy of life, means to find and reach our individual as well as our joint destinations.

"What's on yer mind, Son?"

A giant police officer sank his Irish blue eyes into mine at the corner of Wall and Broadway in New York City. Fascinated, I had watched him directing traffic. I wanted to ask him a question but felt this imposing figure must not be bothered with the trivial and embarrassing fact that I, who had safely traversed notorious Antwerp, Istanbul and Calcutta waterfronts and had a direct pipeline to Planet Earth herself had now been robbed in plain daylight by these alert New Yorkers.

"Hm – ah – I just wanted to know where the police station is."

"What ye want that for? Want to be locked up do ye? I can do that for ye…"

"Just wanted to report that my wallet was picked."

"What shame, and how much was there in it?"

"Only twenty dollars, but this wallet was an heirloom and…."

"Twenty dollars! By gar, I could use that! Now look here, sonny, turn left on the second stop light and there she be. Ye can't miss it. Now take care so they don't kidnap you!"

Wow! My sober no-nonsense friend obviously saw right through all my quirks or pretendings or daydreamings about a pipeline to a personal Planet Earth and was nevertheless willing to hear me out and see if there might be some gold nuggets in there among the junk jewelry. In a while we shall add some verdicts from energy moguls, economists and even crack psychiatrists.

Planet Earth is a plump bundle of energies which she received and continues to receive from her father, the Sun. Even oil and nuclear power are indirectly sun-generated. Another and more benign form of energy is the difference in temperature between ocean surface and deeper layers. It can be tapped easily. Engineers have looked greedily at this stored wealth since 1880. In 1927 George Claude built a machine for tapping it. In 1942 the French Government began building a plant at Abidjan, West Africa, proudly showing a naïve, admiring American all their tricks. I hurried back to the States, bounced eagerly into a Government research laboratory.

A bewhiskered moon face was sent out to see me. He asked the strangest questions.

"What kind of engineer are you?" I ventured.

He shook his head, "You are talking to a psychiatrist!"

I escaped before the men in white coats arrived and steered for the National Bureau of Standards. There a crowd of youngsters pushed their slide rules, "It won't work with those low pressures."

A greyhaired gentleman had been puttering in the background. Now he produced two gleaming gadgets, one running a tinfoil turbine, the other converting salt water to fresh. He was James Hoffman, head of the Chemical Division. His gadgets were hauled into Congress, in support of hearings on the formation of a Saline Water Office in the Department of the Interior. The Saline Water Office was voted in.

Onward, brave warrior, to the University of California and its wizard, Professor Everett D. Howe. We built plants in three sizes, ran an old General Electric aircraft turbine which we had dug out of the hallowed university grounds – so fast that a bearing loosened and broke through two inches of protective wood and missed by an inch the left ear of an Austrian physicist who had gloomily predicted that "nothing would happen at these low temperatures!" he tramped on, muttering, "You did it on purpose!"

We invited key state senators to watch our demonstration run.

Professor Howe gave the signal and I opened the critical valves. A
murky mixture of oil and hot water rained down on the senators'
flashy spring outfits. This flop not withstanding, our technical results were good enough to warrant building large practical plants
for energy production and desalting of ocean water. If 'the market
place' had responded, we would not today have had to worry about
energy shortage, oil pollution, nuclear hazards.

Sky-rocketing oil prices in the early seventies brought the Thermal Difference Energy system to life again. In May, 1975, sponsored
by the National Science Foundation, five hundred technical people
from industry and universities all across the nation tore into all aspects of this energy system, one hundred and thirty of us being assigned to committees working out heat exchangers, ocean platforms,
cold water pipelines, moorings, laws of the sea and general overview
and cost estimate.

The dominating figure on this latter committee was Professor
William E. Heronemus of the University of Massachusetts, who
wrote in his spring 1975 report:

> "Any competent man with a broad-gauge industrial sense
> of what can be achieved by 1975 U.S. industry using the ma-
> terials, energy and financial base available for the next three
> decades, will agree that OTECS (Ocean Thermal Differ-
> ence Energy Conversion System) is that which could be done
> best. The results of this study can be summarized as follows:
> Enough has been done now by many to guarantee that OTECS
> is technically feasible. There are clear-cut pathways to make
> OTECS economically preferable, not just feasible or competitive.
> Large-scale development, acquisition and deployment of OTECS
> would be almost identical to the World War II shipbuilding ef-
> fort. The world has never seen another industrial effort so easy
> to get started and so capable of producing prodigious numbers
> of high-class products. This economy could flood the world with
> OTECS if there were simply a desire to do so, and the effort would
> spread from the waterfront back into every portion of the indus-
> trialized hinterland like the wildfire of prosperity, if we so desire."

A later report from two of our largest technical undertakings, Lockheed and Bechtel, states:

"This report has established that the use of Ocean Thermal Power is technologically feasible at an acceptable cost and at an operating cost comparable to existing and proposed power plants. Half of the Earth's surface in the tropical zone consists of ocean suitable for OTEC operation. The basic energy source, the sun, will continue whether man makes use of it or not. The environmental consequences are believed to be almost wholly benign for any foreseeable level of exploitation. The implementation of the demonstration OTEC plant and the subsequent authorization for chains of production plants would make a major contribution to solving the energy crisis and should become an urgent national priority for the United States."

When could power for such plants be available for use? "By 1981 practical power plants could be available and by 1985 hundreds of them could be producing power, hydrogen, ammonia, substituting natural gas, alcohol and gasoline," writes Dr. David F. Mayer of the University of New Orleans in his proposal submitted to Mr. Schlesinger of the White House. There is complete agreement between Dr. Mayer, Dr. Clarence Zener of the Carnegie-Mellon University, an old OTEC Hand, Mr. Anderson of Sea Solar Power who built the first closed cycle OTEC plant and the 81-year-old author who pioneered this work in the US in the fifties. The condition, of course, is that the nation, or its leaders, make a decision now.

"The basic concept in this program," writes Dr. Mayer, "is that as much work as possible is to be done concurrently rather than serially. Also, when an overall plant design is undertaken, the engineers immediately address the essential questions which must be answered to design the plant. Otherwise, as in the present case, there is a great deal of disoriented research on relatively insignificant problems which have little bearing upon the actual plant design."

Dr. Mayer's proposal, named *Parallel-Telescopic Development Program for Ocean Thermal Energy Conversion* calls for immediate overall design of six configurations, followed by construction of the three considered best, all in the course of four years. Total estimated cost, half a billion dollars, is less than the cost of one week's imported oil.

Dr. Mayer's plan would fit in well as a beginning of a longer-range proposal by the University of Massachusetts at Amherst. On April 21, 1977, Professor William E. Heronemus, as part of the Chancellor's Lecture Series, summarized our energy policy up to that date, then proposed a combination of wind energy and related energy systems, including large Ocean Thermal Energy Conversion Plants, twelve for suitable and needy foreign countries plus an unspecified amount for national use, all within a budget starting at two billion dollars a year and gradually expanding. Compared to presently used or planned energy facilities, this would cost less; additionally, no pollution, no poisonous waste, no running out of resources, a more favorable energy balance and a simple, easily manufactured labor-intensive technology, offering more ample and more balanced employment for less funds.

Five hundred top technical people have worked with this power system and know what it can do. Why have practical plants not been built already? Decisions are made by people who have not worked on this system, yet seem to believe they must personally test all "Technological Alternatives." They cannot. There are too many. The solution is simple: if one reputable industrial firm or one major university investigates a solution and finds it valid, this is worth more than the opinion of any number of Government men who have not done penetrating research in the area. When not just one but many of our larger firms and seven major universities agree on a solution, no sane man would need more "proof". Can a President, a cabinet, or the Congressional Office of Technological Assessment add to the understanding provided by these firms and universities? Can we now today reclaim our enterprising spirit of centuries and build an ample, secure energy system?

The pre-OTECS energy and desalination research at the University of California in the fifties received funds in spurts. In between the spurts I worked at the Los Angeles campus on pollution control and treatment of sewer outflow. We found about that time that most of the local pollution was caused by cars and oil refineries. The prospects of new energy sources were therefore well received in engineering and science circles. However, we calculated that the particular type Los Angeles pollution could be eased by leading the pollutants long natural canyons up to high-placed giant incinerators that would push the smog up through the inversion layers. We had the support of eager young meteorologists but an older variety scolded us severely for "trying to interfere with nature's mightiest forces." One of the younger meteorologists replied, "But sir, these climatic forces appear so mighty simply because they consist of such an outrageously large number of tiny forces with which, incidentally, man has interfered all along. What we now wanted to do was simply interfere with the interferences."

Our further research in this direction was never funded. Pollution control was left in charge of a police officer who swore he would lick the problem by getting tougher.

I rounded out my activity at the University by telling the young, and sometimes the old, what fifty years around the world had taught me about young and old minds, about hate and love, women and men, drugs and drinks and the beautiful life.

All this time Dick West Brooks, President of the Chemiton Corporation near Seattle, had been thinking and talking about another side of the pollution problem, expressed in an interview in the *Seattle Post Intelligencer* on April 2, 1970. Urgently needed chemicals and even food could be extracted from pollutants, he explained. The entire pollution control could be made a profitable business rather than a costly public venture. But he or his company could not realize this all alone. He had found no response in quarters that could make his plans a reality. "The problems are totally solvable, but it is so difficult to get the idea across."

A recent event illustrates this. Thirty-six million tons of sugar cane stalks and leaves are annually burned, polluting the air. The Louisiana State University researchers produced single-cell protein from this waste by grinding, chemical digestion and fermentation. An independent study by Bechtel International found that the process could become economically competitive. A demonstration plant was then called for but, "no sugar producer is at this time willing to fund such a costly venture."

What is the matter with us? Why should a private company be expected to finance such nationally essential research? Don't we have a government any longer? Where has the Federal Solid Waste Management gone?

A broader question: Why do we so continually and steadfastly miss the boat? Psychologist Colonel James Mrazek writes in the *Air University Press*, "Analytical, logical thinking (the only process so far allowed in education and acquiring of knowledge, from the first grade through the highest university degrees) – moves laboriously, a step at a time, sorting, relating, finally concluding. Another type is the intuitive flash-of-insight process. A person encountering an idea all of a sudden is surprised by it. There is no explanation for it." Colonel Mrazek sees this 'creative intuition' not just as an adjunct to the mind but as its main process through which scientific discoveries are usually made, military battles planned and won.

This intuitive process of mind activity has been known to Western psychiatry for decades – to Eastern Yogis and Sufis a little longer. *The Ishopanishad*, dug out of India's distant past, has a word about the type of 'knowledge' acquired exclusively through our analytical-deductive method, "those devoted to illusion enter blind darkness. Into greater darkness enter those who are solely attached to – 'knowledge'."

Mankind is not made up of psychiatrists and yogis. Glenn Williams is a simple Honeywell engineer. He was talking to us even simpler Navy engineers about statistics and probability. One in the audience asked him a question. The words made no sense. Glenn looked blank for a moment, then his face lit up. He answered what the questioner had meant to ask but never did ask. I saw Glenn after the lecture, "You didn't understand that questioner's words any more than we did. Can you read minds?" He smiled, "Sometimes we tune in."

The United States army has for years maintained an office looking for a breakthrough in this art, so intelligence agents can listen in on enemy planning, in a neighboring house – or thousands of miles away. When I last looked up the Colonel in charge, I found him still waiting for that breakthrough. Why not hire a yogi? Yogis, the substantial kind, are not usually for hire by combatants, which an early military man, Alexander ('the Great') found out quite some time ago.

These explosive matters will be further detonated in coming chapters. It would be technically unfeasible and strategically outrageous to develop the full impact here and now. I lived in this kind of world from as far back as I can remember, unruffled by snow, ice, ski slopes, engineering and a stern Lutheran environment.

I wish everybody including business executives and public officials could share with me this bewitched and magic world where problems are solved as they arise, be it in the world of seers or in the land of doers.

Professor William E. Heronemus of the University of Massachusetts, for example, and Lloyd Trimble of the Lockheed Missile and Space Company know what to do about energy. John H.G. Pierson, John Philip Wernette, Leon Keyserling see and know economics. Barry Commoner, Dick West Brooks, Buckminster Fuller know the secret of ecology.

Buckminster Fuller said in *Life Magazine* in 1972, "Pollution is nothing but resources that have not been harvested. We allow them to disperse because we have been ignorant of their value. The government could trap the pollutants in the stacks and spillages and get back more money than this would cost out of the stockpiles chemicals they'd be collecting."

The government could do it or Dick West Brooks could do it privately, probably better. Give such men money to start the ball rolling. Nothing was ever achieved without trying, without providing risk capital. Can we afford it?

The following chapters are intended to show that we can and that our general economy and our life styles would benefit. Clearly, it has to be tackled bit by bit, in smooth cooperation between business and government.

What is our environment? Perhaps a better question: What is not? Our shirts, our pants and shoes are our environment. Our homes, lawns and neighbors. Our neighbor's cat confronted by our own dog. The trees and birds. The paved streets. The oozing cars. The sputtering motorcycles. A daughter's friend who brings reefers and wants her to join and not be chicken. The policeman on your beat. The sheriff's car driving by. The fire engine. The newspaper and tv, the school and the university. The internal revenue. The doc and the dentist. The hospital and health insurance. The wife, the mother, son, daughter. The daughter's baby. The factory. The gas station. The office. The bed and the clean sheets. The Laundromat. The heater, the range, the refrigerator and the air conditioner. The car, the bus, the D-11 and the 747. The air, the tap water, the swimming pool, the ocean. Your body. Your mind. The neighborhood and the next town. The country. Spaceship Earth. The moon, the sun, the galaxy, the universes.

All this gives us our comfort, our tension, our fear and irritation, our health and disease, our development and retardation, our joy and sorrow. The art of ecology is to turn the environment to joy, not sorrow. There are two points of attack: changing the part of the environment that does not make us happy, and changing our own physical stamina to meet and cope with whatever environment is encountered. To accept one of these paths and exclude the other would not do. We need to do both, continually, while avoiding exaggerations.

Plans may be magnificently large, investments modest, at least before we know more about ends and means. To present a huge budget and say: this is what ecology costs, is worse than useless. We don't know enough even of the smallest areas of interest. Take the matter of power plants and their pollutants, for example. There is a matter of choice. There are types of power plants that produce no pollutants. They may be developed through the same kind of enthusiastic drive that landed men on the moon. Following chapters will show that we have more chance of success with such development than seemed possible with the space ventures when these first started.

Similarly, one may dispose of automobile exhaust by using power sources other than gasoline, such as non-polluting fuels, steam electricity, electronic beams. Many of these are already practical and in use, even though gas engine manufacturers advertise their doubts. Their world is at stake, why shouldn't they fight? They provide us with useful second thoughts. But if we let those men or women make the final decision for us, we will be out of luck.

In the general power picture, too, those who sell oil or gas or atomic fission or any other procedure, will insist, and perhaps believe that their own way is the best, if not the only way. We users will have to do better than take their word for ultimate truth.

So just because of this multiple choice before us, any calculation of total cost now is premature. But some day some cost estimate will have to be made. Before that day we ought to understand what money is, how money comes into being, what money we now have or can have, what to be frightened of and not to be frightened of. One who refuses to know about money or one who thinks he knows more than enough and more than this author knows, may skip the next chapter. Once this author also did not know. I shall lead the reader through some of the same stages through which I have passed. Those who read with few prejudices and even those who are full of them may come out refreshed.

MONEY, MONEY, MONEY

I came to Turkey as an engineer in the twenties. The Adana plains were ideal for cotton raising. The Turkish farmers knowledgeable and hard-working. There was an urgent need for cotton all over Eastern Europe and Asia Minor. This would seem a beautiful and secure business for any bank or combine of banks to finance, but those approached had other commitments (maybe race tracks or circus performances?) so the plains remained unproductive and people had to go without clothes.

Then a new regime took over the country, "The Young Turks." Their new Secretary of the Treasury, Saraguglu Shukri Bey, consulted with me who advised him there was no reason why the Government itself couldn't act as banks do, and provide the money, if operating as responsibly as banks do, or preferably, a bit more responsibly. So the Adana Plains were cottonized, everybody profited and my career as an economist, my third career, had been launched.

I wrote books, I helped run a new Scandinavian bank, supported by merchants and the Government, to help relieve the depression haunting us at that time. After World War II, when the Nazi occupiers had crushed the Norwegian economy by reckless printing of money in the face of dwindling supplies, I was named to a commission to repair the damage. This caused our senior economist Wilhel Keilhau to remark, "If that idiot Bryn is going to serve, I quit!" To which Premier Nygaardsvoll responded, "Good, Dr. Keilhau, that rids us of you."

I rejected the commission's report and wrote my own. This *Minority Report* was later accepted by the Norwegian congress. Luckily a better man than I, outstanding Dr. Ragnar Frisch, implemented that policy, while I returned to the United States.

In the thirties I had talked to American bankers who gently comforted me, "Yes, yes, Bryn, we must certainly do something in line with your ideas, though not yet, maybe in twenty-five years."

So, twenty-five years later I turned to ingenious Dr. Seymour Harris, Senior Advisor to the Secretary of the Treasury under Kennedy and Johnson, and caused him to miss important appointments while he listened, contemplatively. Loveable Arthur Schlesinger even missed luscious luncheons. I went to Tunisia, heading a United Nations mission as Economist-Ingenieur, to bring relief to Southern Tunisia. When I came back, a slam bang meeting was planned, to turn this nation into a full employment paradise – when tragedy struck. At present Yale's Dr. John H.G. Pierson, Dr. Leon Keyserling of Truman fame and Dr. John Philip Wernette are working for the same policy of full employment, on only slightly different patterns. Through these and some others' voices and language patterns we shall now look at money, what it is, what we can and cannot afford.

In *The Control of Business Cycles* (Rinehart 1940), Harvard's John Philip Wernette defines money:

> "If we were to attempt to be precise we should probably end with an array of definitions of what is never-money, sometimes-money, always-money, legal-tender-always money, non-legal-tender-often money and many others. The purest form of money is Bank deposits. They have no physical substance but are mere figures in books. Pocket money can be used to make jewelry, to fill teeth or paper walls. Not so bank money. It is generalized purchasing power and cannot be used for any other purpose."

Dr. Wernette goes on to explain how, when a bank grants a loan, new money is created throughout the banking system. This touchy subject has been discussed as long as this republic has been in existence. Some refuse to believe it. Others think it should be discussed only among bankers, not by the general public. Ben Franklin, Abraham Lincoln and Mariner Eccles, Chairman of the Federal Reserve Bank under Truman, all elaborated forcefully on the subject. Dr. Wernette asks, "Is there any reason to suppose that the amount of money so created – i.e. the amount that borrowers wish to borrow from banks and the amount that banks are willing to lend, will turn out to be equal to the amount of money that the people of this country wish to hold, and equal to the right amount for functioning of the economic system?"

In his later book, *Financing Full Employment* (Harvard University Press 1945) Dr. Wernette proposes, "Control of the total amount of money must be assumed by the Government." He envisages a stabilization board for creating or withdrawing money when and as necessary for carrying out the nation's wishes within the limits of manpower, resources and monetary stability.

About the risks involved he writes,

"A do-nothing policy presents the greater dangers. On the one hand we have the danger of mismanagement. On the other we have the danger of mass unemployment and social upheaval. Of the two, the latter is both the greater danger and the most likely to occur. The less risky course is to implement a stabilization program and to make every effort to see that it is skillfully administered."

Anticipating the "bootstrap" gag he writes, "We pull ourselves down into depressions by our hat brims. The only way we ever get out is by lifting ourselves by our bootstraps." For the "Sound Money" boys he has this: "that money is soundest which contributes most to the economic well-being of the country."

Clearly such a stabilization board would handle urgent functions that no private bank could or would touch, but great care must be shown in the composition of such a board. A "pure" government body with only bureaucrats could wreck the economy through ignorance or corruption. A healthy input of private bankers, as in our present Federal Reserve System, seems mandatory. Yale's Dr. John H.G. Pierson seems to me to accomplish a smooth cooperation between private and public interests in his plans for guaranteed full employment. His first book, *Full Employment*, was published by the Yale University Press in 1941. Since then he has been holding responsible positions in the national and international economy and has written three more books and a vast number of articles, some in professional journals (*The American Economic Review*), others for a wider public in *The New York Times, Washington Post, Congressional Record*, etc.

He has stuck with his Full Employment aim as his basic theme. He was one who tried to strengthen the *Employment Act* when Congress was adopting it in 1946 and now he works for the completion of that act so it may achieve what was originally intended. The only condition he would impose on the money managers and other administrators concerned would be a definite level of employment, determined by Congress each year based on the economic situation. Not only is this feasible today, but it would lift our entire economy to a higher and more stable level and achieve more of what we wish our economy to achieve. There would be no leveling-off of income but a general lift at all levels and the awkward matter of welfare would be reduced to manageable levels as employment would be available to all who want jobs.

If the current money managers would not know how to help achieve this, able men waiting on the sidelines would step in assuming, of course, that the citizenry would be sufficiently bright to elect the appropriate teams and not fall for the scare yarns of those who don't understand; who think their position is threatened of who have managed to collect or inherit a few dollars and from then on think the "system" that made this bundle for them must be perfect and any change must have a sinister purpose.

There are even some who philosophize: "A little unemployment is good. It will teach those workers and their unions not to make unreasonable demands" and, so thinking, if they are in key positions they may cause unemployment not merely of the unskilled but, as now, of engineers, scientists, productive minds who contributed toward the standards and the comforts we enjoy.

Insurance is the key to Pierson's plan; insurance not merely of employment but of economic conditions that will provide private enterprise with the desired job opportunities. Since every bit of our economy was established by man, it can and must be controlled by man. The fading faith that our totally man-made economy can run itself "without interference" is referred to in a Pierson letter to the *Congressional Record*, 21 June 1976 Abbreviated:

> "A program of guaranteed full employment based on an "Economic Performance Insurance" approach as recommended by this writer would not only not cause inflation but actually be the best cure for inflation. In briefest summary, under that approach the Federal Government would state in advance each year (a) the minimum and maximum level of employment and (b) so as to assure private business of a continuously adequate total market, the minimum and maximum levels of private consumer spending that would without fail be maintained, for example through changes in consumer taxes and/or transfer payments on some predetermined basis.
>
> The level – or the range – of employment would reflect a decision on what would constitute full employment in a practical sense, with enough but not too much allowance for frictional unemployment. The level or range of consumer spending would be: dollar value of an adequate total market for the nation's output, (Full Employment GNP) at expected price levels minus best-guess amounts for private domestic investment, Government purchases and net exports –all to be adjusted as required during the year.
>
> First, this mechanism would give the economy two good brakes against "demand pull" inflation. Neither employment nor consumer spending would be allowed to rise beyond the ceiling. The dreaded inflationary flood of spending in private markets would not be unleashed. A stand-by tax of a predetermined kind would by advance authorization be activated to hold total consumer spending down. In addition there would be new deterrents to "cost-push" inflation. Business and labor need no more make their sales prices and contracts provide cushions to help tide over bad times to come. Wage and price controls would seldom, if ever, have to be invoked."

Pierson's frequent access to the *Congressional Record* was caused by legislators and economists belatedly seeing 'Performance Insurance' as a condition for success of a full employment policy. Dr. Pierson has worked on these matters for nearly forty years, with constant feedback from his jobs in the business community, with the United States Government and as Science and Economics Advisor to the United Nations.

He writes in the *Congressional Record* March 1, 1972:

"A word is needed about what really is at stake because the arguments over the full employment issue are often pitched on altogether too narrow ground. In briefest summary:

(1) Involuntary unemployment is destructive of personality.

(2) An assurance of continuous prosperity and full employment would weaken the antisocial (usually inflationary) compulsion of business, labor, farmer and other interest groups.

(3) Racial peace seems impossible in this country without universal job opportunity – the present lack of which is also partly responsible for the alienation of youth not to speak of the helpless bitterness of many older people.

(4) Getting rid of poverty would be greatly simplified as a result of the cash-income effects of continuous full employment (more paid labor, less chance of exploiting labor by paying substandard wages.)

(5) The extra wealth (GNP) which would be created under those full-activity conditions – the staggering amounts now wasted through avoidable non-production is needed to help finance programs to meet the problems of the cities, backward rural areas, and the environment, generally including again problems of poverty but not limited to them

(6) Internationally, that extra wealth would confirm our ability to extend more generous aid to the world's less developed countries.

(7) More (and more fundamentally) than that, confidence in our ability to maintain a market adequate for our own full employment prosperity through domestic policy would substan-

tially deflate our fear of imports and our exaggerated preoccu-
pation with export markets and export surpluses; thus it would
enable us to be a "good neighbor" that encourages and helps
the less developed countries to shift "from aid to trade" as they
become ready for it."

The New York Times, 23 January, 1972:

"To attain full employment is not enough; its continuation
has to be assured ... universal opportunity to have a bonafide
life is what America is supposed to be all about ... the vital
amendment to the Full Employment Act (of 1946) would be
this: Congress would be obligated not to rest content with
criticism but to establish final decisions on (a) a full employment target, (b) a consumer spending target consistent with the
recommended government spending program, (c) procedures
for adjusting the job total up or down if the target was being
missed and (d) similarly contingent methods for adjusting consumer spending."

The Honolulu Advertiser, March 27, 1970:

"From the overall economic standpoint guaranteed full employment would make recessions impossible and inflation highly unlikely, paradoxical as that may seem. First, the employment
and consumer spending guarantees would have ceilings as well
as floors to restrain inflation from the side of demand and prevent the price-wage spiral. Secondly, because the government
was offering such guarantees, it would be in a position to persuade business, labor, and farm leaders to agree to follow some
reasonable set of guidelines in establishing their selling prices
so that "cost push" inflation would be restrained too. This is
why the outright guaranteeing of full employment would itself
provide the best cure for inflation."

Dr. Pierson has purchased a piece of eroded, though in his view reclaimable land, on the island of Syros in Greece, where he plants trees, hoping to remake the land into what it was in Greece's heyday; the work of a generalist, symbol of the wholeness of his outlook.

Beside Dr. Wernette's and Dr. Pierson's approaches to continuous full employment we have the more widely known views and ways of Dr. Leon Keyserling, Chairman of the Council of Economic Advisors to President Truman. Dr. Keyserling now operates out of his Washington D.C. office, more alive than ever.

While there will always be periodic changes in business activity it may be maintained at levels that are at all times satisfactory. While some may still dislike their jobs, it is possible, in view of our many urgent options, to offer a choice of several jobs to any applicant. The Kennedy Administration planned a super-survey that would list resources, manpower and potentials and thus put the entire nation to work based, less on workers' previous experience, rather on being taught and trained at work, in the new procedures of coming decades.

A slam-bang meeting, a talkathon of all the groups working on various parts of this jig-saw puzzle was planned when tragedy struck. These workers, who had toiled with financial nightmares in many corners of the world, scattered. The effort collapsed.

These repeated efforts to bring sense and reason into the chaos of our economy has been followed by keen observers abroad, not the least because the whole world's economy and happiness is so closely related to America's. Sir William Beveridge, one of England's clearest and most dedicated thinkers, said on various occasions, "If the United States, possible the only nation able to achieve full employment here and now, would actually accomplish this, she would thereby do more good than by all her aid and all her wars – not only for herself but for all of us."

William Beveridge was a long-time leader of England's Liberal Party and to understand that term in its English version one need perhaps to have lived in England. The English liberals do not base their views or their politics on past or present prophets or theories. They look directly, without noticeable bias, at the complex problems facing us.

This chapter on money, to be complete, needs one more face, a Britisher, A. de V. Leigh, General Secretary of the London Chamber of Commerce for 35 years. He became the behind-the-scenes leader of world trade and lifted the British pound sterling to world prominence.

During World War II he anonymously wrote *A Twentieth Century Economic System* from which is quoted his analysis of the then-existing money or banking system – why it so often failed.

"When the effective demand for goods is increased so suddenly and largely that current production cannot be speeded up to keep pace with it, stocks begin to diminish and prices rise. The first impetus upward may be given by a relatively small increase in demand from the ultimate consumer or it may be due to psychological causes. The upward movement when once started is, however, carried forward by the urgent and largely increased demand of the holder of stocks, whether manufacturer, wholesale dealer or retail trader. His action is based upon fear or greed – the fear that if he does not buy now, prices will go higher and the hope that if he does buy now, prices will rise still higher and he will reap the benefit. The result of his action is, in fact, that of driving prices higher. When prices are falling he holds off the market in fear that if he buys now his competitors will later buy at lower prices and the hope that if he does hold off, prices will go lower. The effect of his holding off is to drive prices lower. This psychological factor could, it is submitted, be reversed and be made to work in favor of stability.

Under the proposed system, trades would know that when prices fell, new purchasing power would speedily be put into the hands of ultimate consumers. The trader, therefore, would rush in to buy before this happened thus helping to bring prices and general business conditions back to stability without intervention. Similarly, when too much purchasing power has forced prices to rise and threatened inflation, the trader would know that action would be taken to contract currency and credit. He would therefore hold off the market waiting for this to happen and by so doing again possibly make official intervention unnecessary. The present psychology would be reversed, with beneficial results."

... From lofty, impersonal circles we now step down to the personal, the painful. If you earn more than the national average, don't you feel awkward? Don't you look at the workload of those who earn less than the average, and see that it often is bigger than yours? What about the President? How can he know and govern people whose income is one tenth of his? Doesn't he feel like cutting his wages down to the average? Or at least in half, to begin with?

It is not a matter of cutting income and spending, but cutting the fat, so we can afford new industry, new energy plants, all that the nation needs, as Governor Edmund G. Brown Jr. of California showed us during a campaign where he beat all comers, for the voters understood also.

THE SUBTLE GAME OF CHOICE

The first Henry Ford gave us the choice of any color on the cars we bought from him, as long as it was pitch black. At that time this was not bad. It saved him money and kept car prices down. Today we may have any color. And we are given choices of any fuel, as long as it is gasoline. True, William Lair has recently traveled around in steam-driven car, and some go electric. But such efforts are screamed down by ingeniously constructed rumors about fabulous costs. Does anybody know the truth?

We are also offered free choice of energy for our homes and factories as long as it is oil, coal or nucleonics.

What about wind, waves, fuel cells, solar batteries or Ocean Thermal Power? Benevolent agencies say: All right, go ahead and research a little, just to calm you down. Then they frighten you with stunning cost pictures, unworldly hurricanes and tornadoes, impenetrable legal jungles, based on things that would either never happen, were resolved decades ago, or cannot be determined except by building prototypes and finding out.

Is there any way out of this maze?

There is, if we accept that a statement by one single reputable firm or university, after years of research by commissioned scientists, counts more than the views of a thousand hired bureaucrats who did not do such research. When the University of California, one of the largest in the world and one of the most sober, had built and tested Ocean Thermal Plants in three sizes in the 1950s, the feasibility and approximate cost were known. When in the 1970s several large industrial firms and seven major universities confirmed these findings, the nation was actually ready for action, first building prototypes to determine cost more narrowly, then roaring into production. Instead, a confused public listens to bland statements by public servants, PH.Ds and Nobel Laureates devoid of relevant wisdom.

While building Ocean Thermal Power plants for immediate needs, research must go on in this and other fields, with the vigor and funds proposed by the researchers themselves, not by budget directors of their political bosses. If this would increase our entire research budget ten times, so much the better. It would still be peanuts compared to other items on the budget, and nothing has been more profitable through the nation's history than funds spent on research.

When research results are in, who'll decide what to do? University and industrial workers, not "officials' without callouses.

Each and every energy system has its warm supporters and virulent antagonists. The supporters usually know more and are more trustworthy than the antagonists. Fusion seems to have a good chance in the nineties, solar space satellites after year 2000, other solar technologies, wind, geothermal, tides, fuel cells and Ocean Thermal Energy plants now, today.

For an overview of the seething issue of nuclear fission we may turn to the pastoral peace of Switzerland, whose nuclear physicist Jean Rossell, recently confessed to *Le Neutron Libere*: "Nuclear energy was born under the aegis of armaments of destruction. Those participating in this application had trouble with their conscience. As soon as a possible utilization that might be termed 'peaceful' was perceived, everybody jumped in…it was psychologically understandable but dangerous… We could have developed solar energy in a much more direct and efficient manner. This, however, was against the immediate economic interests of certain large companies. To defend themselves they presented studies 'proving' that such development, based on solar energy, was not possible."

Professor Rossell goes on to explain how he looked into some of these studies, saw what incomplete or even oblique background material these engineers had been given, what conclusions they had been told to seek. "It is all arranged so that the results will appear negative?" "Exactly." (English translation of the French text)

Many now look expectantly to our huge coal deposits. These are not ready for efficient or safe use yet. Devices are being researched for mining procedures that will not despoil our soil, which is a more critical resource than coal. Method for utilizing the coal directly in the coal beds are being developed. Methods for burning the coal in kore efficient engines are coming. Before these developments are completed, Ocean Thermal Energy can supply what we want – if we are able to ignore the screams of the coal mine owners.

Another good old standby is the dam, with its multipurpose: Hydro-electric power, flood control, sometimes transportation through lakes or canals. After the catastrophe of the Teton Dam all dam builders are now denounced as destructive, ruinous, inhuman. A proposed dam at Auburn in California is seen as a threat to Sacramento's seven hundred thousand people and their capitol. A superdam proposed for Alaska is denounced as a threat to eskimos, caribou, trees and plants.

Dam builders shouldn't be blamed. We don't blame archers or mud slingers of the Middle Ages, neither do we hire them today. Developments have brought to light dangers with the increasing sizes and pressures of modern dams in relation to earthquake trends. Instead of dams we now can build the safer and ecologically more benign Ocean Thermal Plants. We also have, for immediate application at specific sites; wind, tides, solar and geothermal power, and ocean waves, under consideration for a century and now meaningfully designed by the British, who are looking forward to ocean wave energy when their oil supply is depleted. Wind is already the cheapest energy source in certain areas, and is already widely used. Professor William E. Heronemus of the University of Massachusetts proposes extended use of wind in North America and Canada, suggesting relevant equipment.

Ocean Thermal Energy is the most universal alternative and could supply all the energy needed everywhere, if we so wish. The actual equipment has to be located in or near tropical waters at this time, though polar installations are on the horizon. From the tropical locations energy can be shipped to any destination in the form of hydrogen, ammonia or other products. Detailed plans have been worked out at the John Hopkins University, the University of Massachusetts and several industrial firms.

The site and local conditions will determine what type of OTEC plant is suitable: Land-based, or ocean-based, the latter placed solidly on the bottom, or anchored, or free-moving. Each of these may either be open-cycle or closed-cycle. In the open-cycle, the sea water itself is the working fluid and is evaporated, under vacuum. The vapor runs the turbine and is condensed as desalted water and can be siphoned off as fresh water. Closed cycle means that the ocean water is used to heat a "working fluid" (ammonia, propane, refrigerant etc.) and the working fluid vapor then runs the turbine, is condensed and used again. When a French, an Italian and an American scientist first thought of ocean thermal power in 1881, the Frenchman, d'Arsonval, proposed a closed cycle. The French engineer George Claude, who built the first plant in 1929, considered the closed cycle cumbersome and inefficient and built an open cycle system. The University of California, who wanted desalted water from the plant as much as power, also built the open cycle. Against many solemn warnings, a current General Electric aircraft turbine ran rather well, just according to computations, on this low pressure water vapor. The University of California plant was not placed in the ocean, as George Claude's plant finally was. After two monumental mishaps –he lost his cold water pipeline two times in succession – George Claude managed to build and run a complete OTEC plant with a third pipeline. He used only his own money and some friends' so could not afford a proper turbine, but his cold water pipeline was wide enough to demonstrate the low friction losses required and the plant was a worthy forerunner of practical commercial plants.

During the 1940s the French built and tested a full-size cold water pipeline in Abidjan, West Africa, and subjected it to corrosion and biofouling research. They found the problems completely manageable. Then political changes, government emphasis on nuclear power, prevented further building of the carefully designed plant, to the deep disappointment of the distinguished French OTEC proponents, Chr. Beau, retired head of French public works and involved in a great many private enterprises – and many others. The reports in some American papers about adverse results from running of the plant have no basis whatever.

In the US, besides the University of California's plants, Anderson built one closed cycle plant. Other firms and universities worked out detailed plans and cost estimates partly through computer technology. Today Lockheed and TRW are planning pilot plants on ERDA contracts.

Meanwhile other solar systems should be more fully researched. Only a full-time researcher in any system knows how much should be spent and where. Charles Greely Abbot, for example, over a hundred years old, claims the only thing keeping him alive is his yearning to complete a device yielding electric power directly from sun power. Untold numbers are building solar batteries in sun-rich areas and two East Indians have figured electrification costs will be cut in half over a 30 year period by this method. The Rayethon company plans solar batteries in space for collecting sun power and beaming it to Earth, claiming an overall efficiency of 60% from space solar cell to Earth DC power by microwave. About this plan Harvard Nobel Prize winner Hans Bethe said it "made no sense whatever." One of 33 scientists still faithful to nuclear fission, against 2,500 not so faithful, Dr. Hans Bethe joins Harvard physics professor Harold Wilson who back in 1972 told us "Solar power sources are still only dreams."

While our most esteemed halls of learning have thus entered the dream world, younger upstarts, such as the University of Florida, now heat and cool homes, run cars, buses and repair shops on solar energy, jointly with Sweden, Russia, India, Japan, East and West Germany and England.

The plum of the solars, the Ocean Thermal system, is particularly well suited to initiate a full employment policy as advocated by John H.G. Pierson for 37 years, Yale economist of United Nations fame, and Leon Keyserling, Truman's wonder boy, still very active, and John Philip Wernette, author of *Financing Full Employment*, Harvard University Press, 1945. From shipyards to aluminum factories to the chemical industries – all would be boosted by the Ocean Thermal Power system.

Having entered the field of economics and sociology, it is proper now to present Dr. Howard T. Odum of the University of Florida, who views "a single system of energy, ecology and economics" instead of "the confusion that comes from isolating variables in a tunnel-vision thinking." He sees us running out of energy sooner than we expect, since current estimates forget about the energy used up providing and distributing our sources. So he wants to change our energy-surplus-conditioned growth period to a no-growth period. He thinks this can be achieved along with improved distribution and a more satisfactory life style. He sees the present exponential growth as just one phase in a long history of successive changes between growth and "steady state", back and forth. He sees our present continued increase in energy use in the face of scarcity of present energy sources and rising prices as a contributing cause to our inflation – in step with many outstanding economists. For all this, Dr. Odum deserves the nation's gratitude.

This brilliant work has been marred by the two brothers Eugene and Howard Odum's brand of "Energy Analysis." There are any number of methods for Energy Analysis, some of them lucidly portrayed in a feud in *Science*, 15 April 1977. All the listed methods, including Dr. Odum's, try to make valid evaluations of technical sytems regardless of locations and other circumstances. As an example, solar heating panels as well as solar distillation equipment are the cheapest, most efficient method in some cases or areas, and in others, no more than a mile away: the worst, most expensive, most inefficient. None of the Energy Analysis methods reveal this. Ocean Thermal Energy Conversion, a panacea for the whole world, turns out a loser in most energy analysis methods, including Dr. Odum's.

What, in essence, is an "Energy Analysis"? It signifies a wish to run away from any careful technical evaluation and make do with a simple graph or formula, like the economists' "Philips Curve". And, alas, many of our leading professionals are addicts. Our Congress hs recently established "Office for Technological Assessments", OTA for short. On its advisory board sits Dr. Howard Odum, whose Energy Analysis is welcome relief for harassed staffers who have 7,000 pages thrown at them with the request to make an evaluation at noon. A distinguished visitor from OTA to our fourth OTEC workshop in New Orleans 22-24 March 1977 generously though unwittingly provided the demonstration.

Worthy North Americans connect the word energy with only one object: their beloved vehicle. Their prancing, gas-guzzling horses have served them well with only occasional visits to pumps. The feed prices have been advancing at acceptable rates.

What now? The news tells us one moment we'll be out of gas tomorrow, the next moment: only after a hundred years. And the frowning ecologists? Are the exhaust gases killing us or just giving us a sneeze? May we look forward to different fuels tomorrow? Or only in a thousand years?

Thousands of cars, trucks and uses are running without gasoline here in America today. We can switch to any or all of these energy sources any time we want. William Lair, the avionics innovator, has been running around the country in buses and cars powered by steam, some of them given him by General Motors. High cost is rumored – probably computed by our Energy Analysts.

The University of Florida has projected solar power into adjusted car engines. Florida's Mike Lameyer has an engine run by an electromagnetic device plus a battery. Small, powerful batteries have been released by the military for civilian use. Utah's Roger Billings burns hydrogen. It is believed pollution-free. Hydrogen is cheap and plentiful but storing it in the car has been a problem. Billings has found a way to do this. Major car manufacturers support him telling him that the adjustments required for a shift to his engine is less than the yearly change from one model to another.

Eric Cottell in New York puts water in the gas by an ultrasonic reactor and thus doubles gas mileage and drastically reduces pollution. M.V. Hill of Illinois has an air pressure engine. Meyer Steinberg of the Brookhaven National laboratories in Upton, N.Y. would fuel engines with Methanol, produced from air and water. Bill Gray of Texas has a vapor generator in which the water and the combustion products are in the same space. John V. Ecklin of Virgin has a patent on a prime mover consisting of six 15pound magnets and springs, tensed by the magnets and released, thus driving a shaft. This is now being tested by the National Bureau of Standards. The Highway Aircraft Corporation of Sidney, Nebraska, that has built an entirely new streamlined car, hopes to drive it by what is called a "Thermonuclear Engine" by its inventor, Joseph Pepp. This would be hard to accept for the plasma physicists who think of a fusion solution by the year 2000, but since Dr. Bogdan Magdich has come forward with a different fusion process which he hopes to see operating in six years, the idea of a car engine of this type may be just the thing.

However, if such an engine will be offered, the government may not approve it for cars in view of the present grave doubts about anything nuclear.

The only one of the new engines that has won the approval and support of Detroit is the hydrogen engine. This does not necessarily mean that it is the best but that it can more easily be adapted to presently available tooling. It must be recognized that the huge investments in the auto industry's plants and tools make it impossible for these companies to accept technologies unsuited to their existing equipment. If any executive tried, he would be replaced. These most powerful sectors within the American economy therefore, do not have "free enterprise" and do not want it. Neither does the Government, which is supported by, and therefore must support, this sector. Free enterprise only exists among newer and smaller companies, and the freedom of these enterprises is constantly threatened by the powerful unfree sector and by "its" government. If we cannot free ourselves from this vicious circle, important technological improvement will come from abroad. To some extent this has already started.

If some elected officials would run without accepting or spending a cent, or perhaps limit their intake to one dollar per citizen or corporation, our government might again be on the side of free enterprise. Newsmen would come flocking to such unusual candidates. They would not hunt, like the tigers. They would wait for the food to come to them, like snakes. They might well succeed beyond any heavily supported candidate.

In December, 1959, R.G.LeVaux was invited by Dr. Leslie A. Chamber, representing the West Coast Petroleum Companies, to propose implementation of Dr. Carl Page's pollution control devices which contrary to present devices, decreased fuel consumption. They had already been tested to everybody's satisfaction. There never was any answer. Meanwhile, Japan, with its lesser investments, boldly stepped ahead with effective pollution control, better mileage.

From a general health point of view there is hardly any reason to worry about the one year extension of the original 1975 pollution control target the industry achieved, but along with this a jockeying for certain procedures in preference of others is going on. The National Academy of Sciences summarizes the situation thus,

> "The system now most likely to be available in 1976 in the greatest number, the dual catalyst system, is the most disadvantageous with respect to first cost, fuel economy, maintainability and durability. On the other hand, the most promising system, the carbureted stratified charge engine, which may not be available in a very large number in 1976, is superior in all these categories."

The Federation of American Scientists expresses the same view in even stronger words.

There is one bright star in this somewhat muddled sky: the experimental steam-driven cars and buses our incomparable William Lear is playing with, were placed in his hands by General Motors. Is it possible that the auto industry's silence and delays were caused by a genuine hope for an early other-than-gas engine? A Florida company, in addition, has a steam engine running on liquids other than water. And we have the further prospects of solar power and fusion power, both of which could be beamed into suitable automobile engines.

Along with the battle for cleaner engines now looms another crisis: the rising demand for ousting the automobile from large areas so far dominated by it. In the fall of 1953 a personable young Ph.D. came to us at the University of California at Berkeley from the University of Chicago. Ralph Meyer was deep in the study of management: city, state, federal. He wanted to see the thermal recycling plant we had built to convert sea water to fresh. In this plant the power for pumping water etc. is supplied by a steam turbine placed between an evaporator in which surface water in the sea evaporates without heating, under vacuum at which boiling point corresponds to surface temperature – and a condenser whose cooling water is pumped up from deeper, colder layers. Thus the plant operates without any power input. The power is the temperature difference between surface and deeper layers.

This feature, not the sea water conversion, was what interested Dr. Meyer. This plant, he said, would be a link in a grand recycling process, and assist in the sun's photosynthesis cycling, for a future when such an assist would become required.

We asked him about publications of his work. He placed a fore-finger on his mouth: there would be eruptions, he said, if the public learned about it – future cities of 50 million people into which an automobile would never be permitted to enter. What would our 90 million car owners say to that?

"But there must be some notes or reports among yourselves?"

"Yes, and even in those we never refer to these huge city com-plexes as being in the United States. We pretend we are conducting a study on the future of India."

So planners have been concerned with these matters for at least twenty years, though solutions have so far been considered unaccep-table because of the assumed mood of the American public.

The choice of power solution is basically up to this public with its assumed "mood" it is not up to technology. Technology has many alternative solutions. Technology is servant not dictator. In important matters technologies are not even listened to when they try to give advice.

The choice of when and how to install anti-smog devices on cars or produce other than gas-fired engines and pay for the difference in price if there is any difference – is also up to this public and its mood. Technology has stood ready for at least fourteen years with alternative solutions.

In some other countries the masses and their moods are not so eagerly consulted. They tell me this favors progress because the elite can choose. Which elite? I have met people I would call elite both in the US and in Russia. In neither of these countries were these real elite on top in a position to make choices. In the US these elite troops, though mostly hidden in humble positions, at least have a chance of being heard, even if they risk being fired. In Russia they seem to have no such chance. I am happy I belong to a nation where the mass of the people and their hidden elites have a chance of being heard. Though I would appreciate it if the US, land of the free, would make it possible for the elite to come to the fore and help us make wise choices.

It is not always understood that the top positions in governments and private companies are reached by being average, appealing to the average people who elect or appoint them. This cannot easily be changed but, hopefully, it may give way to the understanding that we have to seek the elite and bring them forward in the same way that the Romans went out in the fields and to the factories to seek and find their leaders, and as Norwegians do even today. They find a man, or a woman, whose judgments mostly proved to be correct, who is respected as worker and family member, and ask him, or her, to lead or advise. After serious soul searching, the chosen one may accept.

On the other hand, one who comes forward unasked, offering to lead, was regarded rather suspect by the old Romans and is, by Norwegians, today.

Do not a nation or a city have the leaders the citizens deserve? Deceptive delusion. Candidates for leadership are selected for the flimsiest reasons by a handful of people. The general public vote a choice of two such candidates whom they hardly know. They have no practical choice beyond these two. No wonder there is mostly a poor show of voters. This whole procedure may be changed.

A modern leader, has he really any power? A US president, for example, has the power today to run our economy so that any one who wants to work can be immediately employed, at adequate wages. He does not need to know how in detail. He can select, and authorize the right persons, who have shown what can be done. He can do it while retaining all our freedoms, expanding our options, keeping inflation to a minimum. He may penetrate his administration with integrity, honesty, moderation in salary and expense claims.

We are doing well in the US. Obviously we could do much better by electing leaders with adequate knowledge of available alternatives.

ARE YOU EDUCATED?

Is anyone? People who had to leave school early and are secretly yearning for more education look with surprise and disgust at rebel students at our universities. Even pillars of the educational establishment and of the professions are rebels today and increasingly doubt everything they ever learned.

Carl Jung, pioneering psychiatrist, warns that if you really want to help a patient you must first forget all your science and put yourself in that patient's place and frame of mind. Jule Eisenbud, retired professor of psychiatry at Columbia, wrote me he would like, eventually, "to do your kind of freewheeling thinking, though sometimes I wonder if I have sufficiently escaped the gravitational pull, the educated-in, programmed aberrations of my career..."

John Maynard Keynes, the economist, writes in the foreword to his famous book, "I have called this book the General theory of employment interest and money, placing the emphasis on the prefix "general". The object of such a title is to contrast the character of my arguments and conclusions with those of the classical theory of the subject, upon which I was brought up, and which dominates the economic thought ... of the governing and academic classes ... as it has for a hundred years past. I shall argue that the postulates of the classical theory are applicable to a special case only ... the characteristics of this special case happen not to be those of the economic society in which we actually live ..."

Now, forty years later, if you listen closely to such pioneering economists as John H.G. Pierson, Leon Keyserling, John Philip Wernette, you perceive their rising doubt about all current economic theories.

The teachings of science, all branches, are they facts? Or just tentative concepts? Imaginative dreams?

Such questions require volumes for proper treatment. Such volumes are available for earnest seekers. They suggest that if improperly understood and handled, the educational environment can become as crippling as foul air and water. Every mind must be trained, though not necessarily by filling it with the cogitations of past cacographers, elevated to "facts". We may have sciences, and education, when each person's own mind in the present environment is freely developed and respected.

Beside the conscious and weighty inputs of the educating establishment there are the equally influential, effortless inputs, such as walking in the woods or sniffing the Los Angeles "air". Then there is the reaction of the individual to these two types of input. This reaction, say various pundits, depends on either environment or ancestors or both. Does it? Or is each individual a specific and unique piece of life, related to, yet, wonderfully independent of either environment or family trees? If so, our breathless efforts to impose uniform school systems and curriculae common to all may be neither helpful nor even possible. Rather, new and experimental schools and educational systems should all be welcome, not as patterns for future schools, just for variety, meeting the varying needs of more pupils.

And what if teachers were certified and graded on the basis of compassion, interest in children, flexibility in approach, in addition to the new exclusive criteria of mastery or rather memory of the ruminations of past dreamers, which we term science, including the science of pedagogy? At least we may hope.

Many youngsters feel pressed to plan and direct their own education. Should this be allowed? If yes, then when? At which age? The sensitive educator need not ask. He finds out, by trial and error, again and again. He knows there is no general age limit. If children who have this wish are thwarted, and forced through a general education unsuited to their needs, they may drop out, or worse, even though they may be highly talented. There are, of course, extraordinary characters, who are able to stand any treatment, any courses or systems, without visible harm, though they won't be likely to benefit either.

Those who are so eager to set up a fixed, general curriculum, common to all, often talk about a "well-rounded education". Rounded to what? To that many-colored quilt of past dreams and theories that we call our culture? Or to the individual vital mind that seeks to train and improve itself for the tasks of today and tomorrow?

Some fear that the student, left to his own resources, will become egocentric and even develop delusions of grandeur. Many of the dreamers of the past had delusions and were egocentric. Placing these on pedestals before the students have already caused many of the latter to develop that same egocentricity and delusion. To properly explore and train your own mind is one way to avoid those pitfalls. Alert persons do this more or less automatically, if not interfered with.

Some see the salvation of education in the new up-and-coming technology and its application: cassettes and home-package books and movies, portable reader machines, seminars by broadcasts from satellites. Yes, all these gadgets and more coming may help promote individual and private choice, speed and proper feedbacks, with explanations of difficult or missed points by pushing the right buttons. Though these gadgets also could help autocratic governments or private groups to persuade, indoctrinate, brainwash, whip up resentment and hate. The intent and philosophy behind it all are the deciding matters. Again, these machines could be used with advantage in B.F. Skinner's mutual suggestion and feedback system, for character training and improved relations between individuals and nations.

A different kind of technology was initiated by Jacques Ménétrier MD of France, who took on individuals of any age or profession or non-profession, improved their physical and mental health without any educational input, also without drugs or chemicals, simply with catalyzers.

Before tackling Dr. Ménétrier we must consider when, at which age, educational or medical inputs should start, and do start: before school age, certainly; even before birth. It begins, of course, with conception. A few among us remember vaguely an existence in a dark and limited environment, where we were reflecting mind pictures of a loving and nervous mother. May we identify with one of these not-yet-born and call him "I" from now on?... "I" began to feel too closely intertwined with that mother's worries, fears and anger and want to disentangle myself. I begin to develop a mind of my own, gradually growing so strong that it knows it can get – out!

A tremendous thing happened: I squeezed through a tunnel and came out in a wide space. I would kick, scream and suck my thumb. I was lucky. I could suck my own mother for food and so drank in her living strength beside the food and have felt good and safe ever since, with a healthy resistance to bugs and even to silly thoughts. My friends who were fed on substitutes seem to have missed something. Also I never stayed wet or dirty long. She was always there to wash and change the minute something happened. She was my first good environment. She taught me to always respect, protect and integrate with my environment. When occasionally she went off the deep end in a rage or despair, I called her back with my haunting and hoping look. We were both each other's environment. Bit by bit my father joined us as the third.

We browsed through forest and meadows together and city streets. I met other children and sometimes I knew what they thought. When I told this to Mommy she looked serious and uncomfortable so I stopped telling her. From then on I had some trouble in my dreams with strange little people with sofa-pillows for heads and I didn't tell Mommy about these either for I knew she wouldn't understand and couldn't help. One night (I think I must have been eight) the men with the sofa pillow heads didn't chase me any longer; they looked uncertain, so I chased them. They scattered before me like frightened chickens. I was growing up. Instead of being harassed and threatened by my nightly environment, I mastered it.

This was a great comfort, for there were so many things that still harassed me and darkened my days: groups of children who constantly teased me and sometimes beat me and some of the teachers in the school who became big, strong enemies backed up by formidable authority. It didn't seem to help much that some of the teachers became good friends. Mommy told me school was learning about life – both its difficulties and comforts and if I understood it right, school and all other things in life would become useful teachers. I thought that the sofa-pillow-faced-people in my dreams were a good example of how you could learn from life and gradually master the environment and I had fantasies of turning my enemy teachers into friends and perhaps frightening them as I had the sofa-pillow people.

What is left after parental, school and polluted air environment may climb to a zooming executive with ulcers and endless tiredness or drop to a hood slugging a victim with a bicycle chain. A specie of the latter achievement was caught by a gentle Parisian gendarme who did not slug him back nor give him a sermon, nor send him to a home with bars and locks but asked him suavely, "Doc or jail?"

Everybody could choose "doc" in French towns and cities at the time. Jacques Ménétrier – medic, physicist, electronics engineer, author, artist and mathematician saw physical, mental and emotional environments closely related. He made physical "catalyzers" that tore into the entire personality and changed it, permanently. The "delinquent" held a cubic inch of water with a tiny amount of silver, copper, gold or other metal or combination of metals – in his mouth for a few minutes at a time. The quality of the dosage depended upon the doc's diagnosis. After weeks or months the patient's character would have dramatically changed, assuming the doc's choice had been a right one. If not, they would try with another dose.

At the opposite end of the social register were the tired executives who were given exactly the same kind of treatment for their tiredness, listlessness, boredom or near-breakdown.

I had seen the treatment practiced among the high and low through sumptuous offices along the Champs-Élysées and Avenue Kléber and in the alleys and back streets around Bastille and Luxembourg when I met Dr. Ménétrier , the inventor. Unaffected by his great success with his catalzers and the gold medals he had received from America, Germany, Japan, this extraordinary genius was already off on new ventures and would hardly talk about what was now for him past. Now he had been designing a diagnostic gadget whose electrodes were applied to the same points on the body where Chinese medics "apply the needle" in the performance of their art of acupuncture, or, as it was called in ancient times, the art of balancing the forces.

Quite a staccato exchange had developed between the eager inventor and the no-less-eager admirer in their joint pursuit of truth and since the first issue of the new gadget had just arrived the doctor insisted upon covering my humble self with probing electrodes to all possible and impossible parts of me, whereupon he pronounced: "Bryn, my friend, you are thirty-five."

"Not a good guess, Jacques, I am sixty-four."

"Quel canard! Who speaks about the calendar? Medically you are thirty-five. A baby yet! And you had a serious infection at the age of – let me see – about twenty-one which you have overcome."

"Well, what do you know!"

"Not much, really, besides what I have just told you...."

Dr. Ménétrier created a private environment of his own in the form of a loosely-knit group or organization, "Centre International de Recherche Biologique", including not biologists only but scientists in all fields, allover the world. The group has two centers of gravity; one in Geneva, Switzerland, where the headquarters are; one in Paris, France, where Dr. Ménétrier and his closest friends live and meet for breakfast discussions most mornings at seven sharp. Besides these close friends he has more distant friends like myself to whom he once remarked, "I don't know why it is that I understand you even better than my long-time French friends although what you talk is certainly not French!" which portrays a very common sentiment from a monolinguist to a multilinguist.

One member of his group is George Vieliovitch, the mathematician, who helps the US place weather stations and tracking stations at optimum locations. Another is Henri Coanda, a fiercely young octogenarian; professor of physics at the Sorbonne, who in his younger years invented the jet engine and flew the first issue smack into a barn at which occasion US' Vannevar Bush, called (by US) the greatest living scientist, commented that jets would have no future neither in military or civilian aviation. Jets are spiting him today, as very noticeable parts of our environment; visible, audible, and olfactory, for better and for worse. Coanda also headed a study of French bread – why it had constantly deteriorated. The group studied both the soil and the baking process, found fault with both and after long labor managed to produce a loaf of bread as good as ever. The Government, having ordered the study, never applied the solution: Too expensive.

It was in America Coanda had his greater adventure: A contract with the US Air Force on abolishing gravity. If that sounds insane, his various avenues of approach were even more crazy and that was one reason, and a right one, why the Air Force hired him. No American would have dared probe into such outrageous venues of thought as Coanda did – this Frenchman of Bulgarian ancestry.

We prefer to stick to what seems "probable" – which often turns out to be the least probable. The Air Force even kept the matter under wraps, afraid of its public image, and with reason, in the guilt-ridden environment we have baked for ourselves. But this revelation involves no stolen Pentagon papers. Coanda told me, not the Air Force. Coanda even claimed to have – well, not quite abolished gravity but to have reduced it by thirty per cent. An auspicious environment for high and long jumpers? I am as doubtful as Vannevar Bush was about the jet. But I have a reason for bringing it out: our environment will yield its secrets but only through bold experiments.

Bold experiments alone, however, may not yield the secrets of the environment. There was a king, during those old days that were supposed to be so bad (but were they really so bad?) – who thought after a long reign, that he really did not know enough. He wanted more education. So he resigned his kingship in favor of his son, and enrolled himself as a pupil of an old man who was considered very wise. The king arrived with a bit of flair; after all, here was the king himself seeking wisdom. The old man certainly did not receive that kind of pupil every day!

But the old teacher displayed little excitement. He just assigned the King to a room in his ashram no better than the other rooms, then he left him alone, did not even invite him to the classes.

Day after day the king walked around the compound more and

more impatient, muttering to himself. At last, after a week he went up to the teacher again and said that, after all, he had come here for instruction. "Yes, of course," replied the teacher quietly. "By the way, would you care to empty my waste paper basket for me?"

The king frowned, "Hm, well why not?"

On his way to the garbage can the king met a horde of students who came running playfully and happened to bump into him so the papers flew all over the yard.

"You clumsy idiots," hissed the king, "If I had still been a king I would have known how to deal with you." He collected the papers again and went on his way.

More days passed without any teaching. After another week he went up to the teacher again, "Say, my friend, I came here to receive instruction you know."

"Yes, I know. By the way, would you mind emptying my waste paper basket?"

The king picked up the basket, laughing this time, but when he crossed the yard another bunch of pupils came running and bumped into him and again the papers flew all over the yard. This time the king said nothing. He only sent them a furious glance as he gathered up the papers.

Another week went by. Again the king approached the teacher, this time beginning to suspect a plan and purpose for the way he had been treated. He was more polite.

The teacher said, "I am glad to see you, John. Say, could I ask you to empty this waste paper basket for me?"

The king took the basket, was not surprised when a batch of pupils came running, bumped into him, so all the papers flew all over the yard. This time he said nothing, did not even look angry. He chuckled, gathered the papers and went on his way. That afternoon the teacher began his instruction.

Is this in line with B.F. Skinner's positive reinforcers? Or are they negative reinforcers? Adjusted to a special individual? Cruder than Skinner? Or more subtle?

Henry W. worked very patiently on his Ph.D. thesis *The Effect of Intraslide Redundancy on Picture Recognition*. He had to adjust to his professor-judges and forgive their sins, to obtain his degree. Henry happened to be naturally sober and patient, as the king in the story became, but was not at first. For the king, like our modern Presidents, executives, supervisors and 'self-made" boasters had slipped in mind clarity through conceit and listening to yes-men. They had slipped away from the sober patience that enables a person to see and consider other and contrasting viewpoints.

WHAT IS AMERICA ALL ABOUT?

Fishing? To youngsters and oldsters alike, to peasants and some presidents, this is all that really matters. Golf? Mary, Queen of Scots set the example, followed by more presidents and their cronies. Firefighters? Those boys (and girls too, lately) are more serious. Skiers, mountaineers ply their beautiful art and sometimes help win wars in the bargain, the latter a worthy and perhaps exclusive American pastime? Ecology? an ever-greater crowd flock to that banner, some for a hobby, some toiling. What about the older heroes: Democrats? Republicans?

"The American Major Parties," writes Alexander M. Bickel in *The New Republic*, "need to be open and substantially overlapping to perform their function. Ideology, clear unity of purpose and interest – these are for factions which do not by any means disband after the election. Parties are for gathering in factions and for holding them together, however tenuously, for a time and for some purpose."

So much for our political environment, which is said to be less and less important in American lives – except for the politicians themselves and possibly for the *New Republic*. We shall look further for what America may be all about.

Could it be work? Its character has changed. There were times when work came as a matter of course to everyone. We did not have to apply for it or seek it. Now we have so structured our social environment that work is now a problem for almost everybody; a tragedy that makes life meaningless, almost unbearable for nine million among us.

John H.G. Pierson expresses it thus, "Involuntary unemployment

is destructive of personality." (*Congressional Record* 1 March 1972.)
John Pierson is a Yale economist. His ideas are not newly thought up.
He has had them and fought for them since 1941. In *The New York
Times*, 23 January, 1972, he expressed it thus: "Universal opportunity
to have a bona fide life is what America is supposed to be all about."

Because Dr. Pierson is an economist whose career has been glued
to the burning problems of domestic as well as international economics, he not only has goals but knows how to attain them. Not
all economists agree, or have that ability, partly because they did not
have Pierson's experience and partly and more generally, because
they are trapped in decorative theories rather than looking at life as
it is. One may have a Ph.D. and even be Chairman of the President's
Economic Council without understanding the flow-charts and potentials of the nation's economy, though some such chairmen have
understood it.

One was Leon H. Keyserling, who served under Truman. Though
limited by conventional barriers all around him and not being allowed to establish permanent improvements beyond his time in office, he nevertheless helped the Truman administration make a better
average record on employment, growth and price stability than any
administration since. The credit goes to the whole administration,
for a lone economist can work out his ideas only in a responsive
environment.

Also, it is not necessary to have passed through a University economic course to see and understand, though it sometimes seems to
help.

So dense is the social environment around us that a group who
worked under the John F. Kennedy Administration for the purpose
of lifting our economy to a level that would offer employment to any
one seeking it (and in the bargain performing some of the urgent
work required in our ghettos, factories and schools and generally en-
rich the nation, both the rich and the poor) – worked without contact
with or benefit from other economists who for a lifetime had been
at it. Some were brought to our attention by alert young college students – those live-wire young who eventually will take over from us
half-sleeping oldsters and build a new America.

If anyone feels uncertain about whether full employment can be successfully practiced in the American economy he should reread the previous chapter on "Money, Money, Money". Whether he does or not, a further trip with John Pierson will be rewarding. One instrument for achieving full employment is already in our possession. *The Employment Act* signed into law by Congress in 1946 – partly, it may be, in consequence of Pierson's books and actions – was a useful beginning which Dr. Pierson now proposes to complete so it will do what was intended. Congress, rather than just making comments on the economy, should each year determine the employment target and consumer spending target and should require the President to act so that those targets are met throughout the course of the year.

If there is any doubt in a citizen's mind that this could effectively be achieved – and with acceptable price stability – his duty to himself and his nation demands that he investigate, study, experience, until he can see the economy as a huge system of flow of food, furniture, housing, defense, foreign aid, money and credit cards.

There are still serious limitations. Not everybody can have the job he most desires; not even always what he may call a satisfactory job, but he can work and make sufficient income for his family. After this has been achieved we may go after finer adjustments, bringing us closer to optimum functioning.

Among other ways to full employment is John Kenneth Galbraith's more or less permanently frozen prices and wages. This would require stronger central control of both economic and social life than we have had so far and besides, would rigidize our technology and delay progress. And there is John Philip Wernette's way, to have a Federal Stabilization Board, which would mean less control of prices and wages than Galbraith's, but would still involve a monetary control for which we may not yet be ready.

Pierson's plan elegantly achieves the same goal by a minimum of control, no more than we have now, and even this minimum would lessen as the method becomes standardized. We would feel more free than now and in addition blessedly liberated from the fears of economic doldrums.

Equalization is not prerequisite. That is neither possible nor really desirable and it would counter the American dream. While the poor would, in general, be lifted above their curse, the rich would generally become richer and the average income soar as taxes diminished with the greater income from which to deduct them.

"The candidates are setting their economic sights too low," wrote Pierson in the *Washington Post* May 14, 1972, during the presidential primaries. This is the tragic story of all of us, the whole nation, in a nutshell. We are setting our sights too low. We don't all have Pierson's rich experience. In addition, those of us who have a job, do we bother too much about the panic of the jobless or the fact that we may be fired next? Above all, do we worry about the even-more-tragic effect on the nation as a whole?

Is this what America is all about?

Perhaps it is too much to expect that people should read books, such as Pierson's, Wernette's and Keyserling's, or even articles. If they did, and came to understand what enormous importance full employment would have, not only for those so employed, but for the whole nation, its sanity, its face to the world and its muscle, its ability to help, and to represent and carry out its convictions, if this would become clear, you might see people reading this stuff rather than the Sunday funnies.

Ed was a top engineer at Boeing. His special project floundered and he was laid off. After three weeks he had a new job, with the government, where I met him. He had never had a day off in his life before. He had never thought about economics. Those three weeks among the jobless made a new man of him. Now he reads books, writes letters, goes to lectures. Now he is angry. I wouldn't be surprised if he remakes all of us one of these days.

Is there anything more important to us than our work environment? The money environment perhaps? Much of the same order and closely linked to the work picture. Or the "mind" environment, subject of the following chapter? This is also closely linked to our work environment.

Our "work environment" is in a sense an alternative to what we now call the "welfare environment". The idea that we have now reached a stage in civilization where many people cannot and should not work but be handed a dole instead, is worse than a death sentence to those involved who are shut off from the sphere of useful activity because of entirely unreliable, often ridiculous intelligence tests, school grades, personnel department idiosyncracies, all of which were disproven by such companies as Lockheed, Bell Telephone and others when they decided, as an experiment, to hire people who had been known as "unemployable." The supervisors who received and tested and employed these people found them quite often superior to the run of the mill of employee who had successfully passed all the "tests".

The welfare protagonists commit another and more serious error. They talk and act as if we have done all we ought to do and now have just to keep the jigsaw puzzle moving along by the easy effort of a few privileged "workers". Actually we have barely begun putting our house in order. We are so far in arrears we cannot even be sure we will succeed. Every hand in the nation, all our bodies, all our minds, are sorely needed. We will have failed as a nation if we are not able very soon to establish a society in which every idle hand can go right to a nearby employment office and have a choice of several regular paying jobs. This is possible. This is necessary. If any one claims it cannot be done, he should stand aside and let others do it.

Shall we just start a count of who could do it? John H.G. Pierson can do it. John Philip Wernette can do it, even though he would do it a bit differently. Leon H. Keyserling, who was head of President Truman's Council of Economic Advisors can do it. Seymour Harris, who was Senior Advisor to the Secretary of the Treasury under Kennedy and Johnson can do it. Paul A. Samuelson, our incomparable economics teacher, Nobel Prize winner, can do it; and many many more, if they are so authorized. Our heralded freedom will be retained and offer more freedom to more people.

When Lockheed and the Bell Telephone Company had such excellent experience with their so-called "unemployables" (people who had never worked before, who had been on welfare for generations, who had police records, some who couldn't even speak English) – why did we not see a major effort afterwards? Instead of finding out why, the entire plan was shelved – to the satisfaction of those who had been afraid their precious employment and management philosophy would collapse. What do we do if this philosophy has marred our lives and endangered our continued existence?

When the job seeker is faced with his fifteen choices at the neighborhood employment office, none of them may be what he most wants or in his "line of work" or reachable from his present home. What then? It should be made clear to him that the nation is in need of people to fill the jobs offered, and he may wish to take on an offered job, temporarily at least, even if it isn't a perfect fit. The employer, or the government, will pay for his moving and relocation. If he sees another job, more in his line, coming up soon, he may of course prefer to stay on, eating into his savings.

In view of the urgent need of workers to do the many jobs now waiting to be tackled, there will necessarily be a certain pressure in the form of much higher wages for any type of work than the level of welfare payments.

Most people would prefer work to welfare anyway, and many today would not mind changing jobs and locations.

The work environment has changed throughout US history. At the time of the Pilgrims any new arrival was a welcome addition to the urgently needed work force. The word got around, increasing numbers of Europeans and Asians, dissatisfied with their lot at home, crowded into America, while Africans were kidnapped and sold as slaves. Human life became cheap. Laws and rules began to limit the opportunities of the immigrants. Employment was made to look like a privilege, obtainable only at the grace and nods of employers. This state of affairs still persists, though it is not reflecting the true needs of our society. We are gradually finding out so many things that need to be done, so many neglected areas that we are beginning to wonder whether we will not need every hand – for a hundred years to come?

The average economist, pursuing only that cluster of subjects that through generations have become arbitrarily added to what is termed the "science of economics", has had difficulty adjusting to this age or seeing its needs. You have to be something of a generalist to do that, mastering at least basics of engineering, farming, physics, medicine and psychiatry. Difficult? Not very. Lots of people today live and breathe in a healthy general knowledge. Unfortunately they don't become famous that way. Their articles are only now beginning to brighten a few forward-looking magazines. The "youth revolution" was in part a reach for essentials. Generalist college students furnished me with many of the facts of this book. At twenty they had more understanding than some teachers at the end of their careers.

For mind is ageless.

And mind is not toothless, and is anything but uniform.

Saadi, a sage who lived in the Middle East some time ago once placed himself in the middle of a busy road, which was a good place from which to advertise one's wares or wisdom at that time, even though it would not be so convenient on Interstate 5 today. He was surrounded by the usual crowd of admiring pupils. Galloping horse-hoofs drummed in their ears and soon a formation of riders appeared. They shouted, peremptorily, "Out of the way, old man, the KING is coming!" As Saadi and his flock moved hurriedly off the road, the sage said, "That's why."

When the riders had passed, the sage went back and sat down in the same spot with his pupils around. Now distinguished courtiers came riding, slowly, stopped their horses and said to the sage, "We think you better move aside, for the king is coming this way." The sage and his flock moved again, less hurriedly than the first time. When the courtiers had passed they moved back to the same spot in the road. The sage said, "That's why."

Now a radiant person came riding along, at a dignified pace and when he saw the sage and his pupils he made a big detour around them, bowed as he passed and saluted the sage courteously. This was the King himself. "That is why," said the sage.

"What did you mean by 'that's why'?" asked a pupil.

"I wanted to point out to you the differences of minds and the subsequent differences of behavior and action," explained the sage. "The fore-riders talked or shouted and acted the best way they knew. They acted as their minds directed. That's why. The suave and polished courtiers with a gentler type of mind acted as was natural to them. The king who knew a brother soul when he met one, acted like a king. That's why. Never judge people from or by your own standards. They act as they must. Never judge at all."

"That namby-pamby!" thought some onlookers, "When he starts serenading, we'll have some fun with him."

Then, as they had rightly presumed, the sage began singing his poems, particularly one about how he saw GOD in every creature, "And whoever comes along, I shall know that it is you!" he hummed and an irreverent bystander interrupted, "But suppose it was an ass?"

"Then," said the sage, bowing serenely to the heckler, "I would know it was YOU."

MAKING CITIES LIVEABLE

A New York University Project for Security Design in Urban Residential Areas under the leadership of Oscar Newman, found, after several years of study, that certain architectural features of residential units, combined with community organization, would render these areas more or less crime-free, enjoyable and suitable for family living and growth, regardless of the size of the city. There is reason, therefore, to change the current concept about the effect of size, as expressed eloquently by Sidney Harris in a recent column: "New York, the archetype American city, where the rich move out, the poor stay in, so schools deteriorate, public service withers, the tax base erodes..." and "in the small towns the young get out, the old stay in." So, he concludes, the solution is clusters of core cities no larger than half a million, with attached satellite towns of twenty-five to a hundred thousand, which, again, would service villages on their perimeters.

This is one solution, while the New York University study indicated that not-too-tall residential units grouped around a court so that all inhabitants have free view to the whole complex and any visitors, will form a natural protection and promote a feeling of community and common interest.

When Ralph Meyer came from the University of Chicago to the Berkeley Campus of the University of California in 1953 to talk confidentially about cities and their problems, he was a link in a long chain of workers who already for decades had broached this vital matter that only recently has evoked the general public.

There is no shortage of solutions. There is a shortage of action.

Why? First the reluctance of Americans, generally, to take part in national planning. Every good business enterprise plans in utmost detail but not so the nation, the biggest business of them all. There is a superstition that national goals take care of themselves through the wonderful free enterprise system. While every one of these free enterprises of which the system is made is carefully planned, the system itself must never be.

The second reason is closely connected with the first: Money. The nation as such has no idea "where the money is to come from" for the formidable task of rebuilding cities and towns to liveble units. For in money matters more than in anything else, the nation as such refuses to plan; refuses even to see or to find out.

There are individuals who see very well "where the money is to come from" as outlined in a previous chapter. Before we make up our minds to listen to these we can have no action. Before we ask every congressman, every governor and every presidential candidate that and how they will release the money, which of the experts they will listen for and follow – before we ask this crucial question of any candidate for public office, we can just kiss a viable America goodbye.

When the required money is made available through the already-established money – and banking system and based on factual national potentials rather than on clashes of contrary concepts and ideas, we may work out details according to any one or a combination of carefully laid plans. Many of these plans call for the communities themselves to take care of their own improvements, moves, rebuildings with guidance and funds provided by governments to the extent desired. For many ghetto-dwellers this will be their first chance at regular paid work and for building their own towns. There will be plans but these may be changed by the local builders (causing delays and sometimes belly laughs?) Belly laughs are good and as to efficiency look who's talking! You-all who wasted billions that would have been produced if you had let the millions of unemployed work. Now is the time for relaxed, unhurried use of every hand. Efficiency we have never had. That may come later.

The factory-built house is made for this self-building system. Just at this time the social and economic resistance to factory housing is dying down. Even builders know it is coming and cannot be stopped and so much new housing is needed that there will be plenty of demand for the older type, too, when our money is made to match potentials.

Some will prefer living in such out-of-sight structures as large domes, geodesic or not, Buckminster-made or not, domes within which space will have a new and exciting meaning. The widespread experimenting in new social forms, together-living in groups or families – have already prepared many for living in open spaces under weather-protecting structures.

As many would move out from crowded ghettos to such new quarters, those remaining, who perhaps prefer the old places, would have room to move around, breathe and begin improving, cleaning, decorating, driven by the undying impulse of humans when vision and a hope are in sight.

Drug addicts, driven to crime for satisfying their need, present such a common sight in cities and towns today that it imposes itself on any urban planner or builder. Yet, this is a much smaller and easier matter than drinking, smoking or overeating and has been nicely solved where people really tried. In England the addicts were given their potion free. Pushers had no business any more and vanished. Some addicts died, a few managed to extricate themselves. In China they used a different method. All pushers were shot, the addicts were kept completely away from the drug. Some died, a few recovered.

If we want to we can get rid of the drug curse and fifty percent of our crime in two years.

While Harris and others dream about new cities not large, not small, other builders are preparing for super metropolitan areas as for example a combined Washington D.C.- Philadelphia-New York-Boston. If not planning for it, it seems to be coming anyway. Can we manage it? Or will it break down as New York did?

Tokyo, a larger city, is in trouble too. People faint from foul air. London, on the other hand, has improved so much in the past twenty years it feels like a new and happier city. Paris is improving; Rome, New Delhi too. It can be done. A city of any size can be handled, managed, improved in every respect. Not by a commission, perhaps, but by people.

How did the foreign cities improve? By American aid?

American aid helped all these nations recover after World War II, a brilliant thing. But the aid was only a tiny part of the total effort required. An almost-incredible will and energy on the part of the people themselves did the trick. And as to the cleaning of the air of those cities as well as the spirit, Americans and their aid had hardly any hand in that. On the other hand, we Americans learned from it. For one thing, in England and France a much larger part of the population understands what money is, how it is created, how it should be administered. They have a harder task than America having less natural resources and smaller populations, though the latter disadvantage is now disappearing through the Common Market.

The Londoner today feels like a new man. He steps out in his park, see the sun and breathes. In 1944 I felt my way along a row of houses on a London street. I couldn't see my feet, only that London "Fog" that we in the US call smog.

The situation in today's American cities and towns, why and how to improve things, may be seen by comparing Bruges and New York.

When you first see the Belgian city of Bruges, a personality leaps out at you. There is history, past and present achievements, grace – all unified. You can see a similar thing in Rockefeller Center when you limit your view to this Center only. But here are only business offices, no residences. It is incomplete as a city unit. In Washington Square you had some years ago a semblance of a unified residence complex. Not many blocks away you see and smell disaster. Rats hustle; garbage dominates.

The Marcus Gerard Foundation watches Bruges with a hawk's eye. If any owner of any house in Bruges lets his home deteriorate he may be forced to sell the house. Nothing is permitted to sink below the standard set. Such "interference with the freedom of the citizen" wouldn't be tolerated in America – not today. Yet, if a man murders by shooting or strangling, we do interfere with his freedom. If he murders by letting his house deteriorate, so babies born there die or become criminals, we don't interfere.

To understand this subtle difference we have to look at the people in Bruges and other European cities who immigrated to the US. We have been told it was the elite. Was it? I am involved myself, an immigrant from Oslo, Norway, so I better be careful.

The immigrants had more means, so they could move. They had a spirit of adventure. Both good things. They also were impatient; not, perhaps, members of the Marcus Gerald Foundation that watched the face of their city with such eagle eyes. They were more insouciant. They went to the US and slapped together a dwelling wherever they pleased: Simple if they were poor; elaborate if they could afford it, with no particular regard for their neighbors or the neighborhood. So their cities became an incoherent mixture of everything and nothing. Above all: Nobody took the responsibility for as much as a whole city or town. Some thoughtful families who had the means took on a certain responsibility of a Rockefeller Center but cared nothing for the adjoining blocks except that they had a pious hope their good example would be contaminating. A scion of the Rockefeller Center family governed New York State. Didn't he feel like using his intelligence and power for making the whole of New York City a well-run Rockefeller Center?

The insouciance of the immigrants does not need to dominate us any longer. Some of our citizens have never been dominated by it. They have been ready to integrate the whole of New York City into a large livable unit. They can do it. So far the insouciants have stayed their hand.

The insouciants' grip on the Federal Government is worse and has more sinister consequences than their grip on New York City. Their aversion to thinking about the whole nation and caring for it as a unit strangles our work, our initiative, the vast innate power of two hundred million units of fantastic potentials. As long as one single man or woman eager to work goes unemployed, even for a week, we are grossly neglecting our duty; devastating minds; crippling the economy.

A striking example of all this is the city of Jerusalem. Well-meaning people tell us it ought to be "internationalized." Teddy Kollek, Mayor of the city of Jerusalem, says in *World Magazine* 7-18-71: "Internationalization hasn't worked anywhere – not in Danzig, not in Trieste." Why? Because the international community and its organs are just as insouciant as the majority of Americans. Israel, on the other hand, has taken Jerusalem under the wings of the whole nation and has done for the city what its own tax structure could never do. A city is part of a nation. If a nation forgets that, some of its cities will die. Not all cities are equipped to finance themselves. The United States has reluctantly realized this and reluctantly paid some bills, though in a haphazard, limping manner that cannot and does not achieve what was vaguely intended.

The editors of *World Magazine*, including Norman Cousins and S. Spencer Grin, have graciously allowed me to quote from Mayor Teddy Kollek's article in which he comes down to the nitty-gritty of race and the housing problem:

> "The worst off are the young people. A boy who comes home from the Army and wants to get married and can't see his way to get an apartment. There are almost no apartments for rent; they are practically all for sale. To some extent, this has driven young people out of Jerusalem.
>
> I think that with some effort we can catch up with our housing problem in five or six years. One advantage is that we have not had the ideal of the melting pot. (hear ye, hear ye!) Certainly not as far as Moslems, Christians and Jews are concerned. No Jew ever wanted to become an Armenian. No Greek wanted to become a Moslem. No Latin ever wanted to become a Jew (notice the delicate touch of the Mayor!) So they have lived in a mosaic rather than a melting pot. And we are keeping this. It makes, I believe, for less tension. It is what people want. In fact, it is a pluralistic society in the best sense of the word. Why change it? These traditions are worth something. The greatest danger to cities is when traditions break up without anything new replacing them."

Mayor Kollek goes on to tell about young students often from the poor districts visiting families in these districts, helping the children with their homework, increasing the feeling of belonging among these students and the families they visit. He mentions "pocket parks" and the importance of keeping them clean and neat, which he observed we don't always do in the US although the idea of pocket parks may have come from the US.

"We haven't learned how to run a city – anywhere!" muses Teddy Kollek and that is the first and important lesson to learn for when you know that, you are willing to try, to experiment, and not start out with the deadening assumption that it can't be done.

LOVER AND BELOVED

The US Navy supplied me with safety shoes that I have worn day in and day out for five years and they are still waterproof and utterly comfortable. I don't even need to shine them. If you look around and are able to spend a little, you may have the coziest environment around your feet here in America.

My daughter bucks the game by not wearing shoes at all, except when she wants to see a movie, and the movie house guardian insists on her wearing shoes. She feels just as cozy in her bare feet as I feel in my old navy shoes. People tell her she will collect all kinds of bugs and diseases running around barefoot. Why? She washes her feet now and then, but who washes their shoes? There may be more chance collecting germs wearing shoes than going barefoot. I approve of my daughter's feet. I approve of people doing their own thing as long as it makes them feel good and they don't bother me. Besides, my daughter is saving me money through her non-shoes, though admittedly, she is taking it out in other areas.

One is reminded of the English and American shoe salesmen who came to an African town, population 100,000.

The Englishman wrote home, by surface mail, "No sales here. Everybody goes barefoot."

The American wired, "Ship, by air, 100,000 pairs."

How can we hope to manage our common environment if we cannot even stand each other's idiosyncracies? When I see a hippie with his hair and beard and patched Nehru jacket, I feel almost naked and certainly awkward in my own 1950 suit and tie and I think he looks fine and hope he can stand me. If he can, I believe we can tackle the environment together.

My daughter says that, walking barefoot, she comes closer to the softest and sweetest environment – the grass and moss and flowers in the patches of nature still available. In between, admittedly, she treads on the dust of streets which is not so invigorating but she wipes it off again when crossing the next patch of greenery.

Is it really true that grass on your feet can do things for you? Not only that but your feet and even your mind can do things for grass too. The effect of your sentiments and thoughts on plants was first measured by the Hindu botanist, Chandra Bose, and now has been confirmed by Western scholars.

When we are inspired by the environment, it is also inspired by us. and what happens when we poison the environment? It, right away, reaches out and poisons us. An old friend who wears tie and a high collar and shined shoes to important conferences spends fifteen minutes every morning walking barefoot on his dewy lawn. Another equally important citizen wades through a shallow strip of the ocean in front of his home every morning and eve. They trot out in their environment and it caresses their feet. And souls?

Divers and swimmers prefer to have the environment caress their entire bodies rather than just the feet. The sun bathers enjoy caresses from another and more fierce part of the environment which may cause them havoc if they don't take precautions and expose themselves gradually.

The answer to the riddle of the environment – could it be love? Meaning that she or he who loves it can coordinate us with it, tame it? But how can you tell the hierarchy of bureaucrats to love our environment before they make plans for it? Members of the hierarchy may still love, at home at least, but the hierarchial system is another matter. For one thing, civil servants at grade nine are so preoccupied about how to make it to eleven, they have hardly time for love, they don't even love that higher grade, it is just that they know they will be considered dumb clucks if they don't make it, and they don't like being considered dumb clucks. I am speaking for myself, of course, nobody else. I was in civil service fourteen years.

The GS nines or elevens may still make plans, but will they be plans of love? Mainly plans they hope their supervisors in the higher grades will approve. The people caught in hierarchies have often lost, not merely love, but the skills and rhythms and good sense of the working stiffs.

Our humble climbers in their awkward hierarchies may still be useful if they know their place and leave the planning and the building to those buck privates who still have love. Offer these lovers modest funds to try out their plans and don't worry if a few don't pan out. There is one thing, at least, we ought to know today: statistics.

Every statistician knows that in every undertaking there are bound to be both failures and successes.

The reason Barry Commoner came closer in his evaluation of and recommendations for our ecology is not that he was a better scholar according to the prescriptions of that game. Nor was his mathematics better than the pallid effort of his opposite numbers. Could it be that he was in love?

Granted that love is the criterion, how can we find out? How can we determine whether a man is in love and so should have funds for his try-outs? Well, a perceptive person knows. In the nationwide organization of aid to small business ventures, some excellent choices were made and funds were lent to ventures that catapulted bold pioneering thrusts into very profitable businesses and enriched our nation in the bargain. These business ventures constitute one important part of our environment. Why wouldn't we be doing as well with the part called ecology?

If we had had a national ecology organization at the time columunist Dan Coughlin brought Dick Wes Brook and his business ideas on ecology to our attention, how could the Federal agents have failed to give Dick a grant sufficient in Dick's own estimation, to try out his plan? We would have been ahead now instead of wallowing in deadwood plans structured by high authority.

In the area of the clothing environment as we move up from shoes to socks, pants or dresses, shirts or blouses, ties, hats, we notice that women in general live longer than men and enjoy life more because they don't bale themselves in clothing that keeps nature's environment out, and don't stick so stubbornly to old customs and standards, but experiment a bit. I am a male myself and ought to know. For sixty years I was a slave of that noose we call a tie because the office custom demanded it and I did not feel I was so good an engineer that I could afford to buck the game and be called a kook or nut. I might even have been booted.

Only when a supervisor three grades above me wrote in the "Plan of the the Day": "Get out those sports shirts, men, and forget your ties," on a warm summer day in Keyport, did I finally break down at the age of sixty-nine, and threw away my tie and moodily contemplated a lifetime of suffering. Who said slavery has been abolished?

Because of this greater sartorial sense of woman they possibly also have a greater sensibility to the environment they are are closer to. So they should have a greater say in ecology, in deciding who are the greatest lovers with the most practical sense who should have the grants to tackle our environment. People who love bulldozers and concrete have for generations played with their bulldozers and concrete and endeared themselves to hierarchies who became colossal planners in bulldozing and concretizing. These may not be the best to handle our environment situation. As an engineer, I know. But their part in the great cycle of nature is minimal. One might say: Bulldozers and concrete won't solve our ecological riddles.

Biologists, medics, botanists, chemists, oceanographers, physicists, geologists, geographers – all these will contribute to the solutions though not in the pay of a civil service hierarchy, a departmental hierarchy, a White House hierarchy or a United Nations hierarchy, for the archangels of these hierarchies with the best of intentions stretch out their authoritative hands and strangle genius. The single entrepreneur who in loving contemplation swallows a dozen sciences and digests and rearranges them into the sharpest tool – he or she is our savior.

Just as a sensible attitude to the shoe-and-clothing environment may indicate love for the entire environment and an ability to deal with it, so also does one's attitude to other parts of the environment, such as neighbors. A person who loves his neighbor ("as thyself") and therefore adjusts to him may be a good worker in all environmental fields while one who thinks his neighbors are the cause of all his troubles may not be so good. There are exceptions, though. One may be terribly biased in one direction yet be very clever in certain fields.

I knew a girl who saw all her difficulties as having been caused by vicious, willful schemers. Yet, she was the greatest cook. I don't think I would have granted her funds to improve the greater environment. But I could trust her come-any-day with my shirts, meals and floors, a not-negligible part of our daily environment. On the other hand, one of my best friends loves his neighbors so much that he has never a cent left in his pocket nor any clothes that he does not wear and often not even a blanket on his bed; certainly not a lawnmower. This man is a famous scholar and teacher of archeology. But would I grant him funds to better our ecology? Perhaps if his wife could handle the funds.

It is this rich variety in our human environment that is so fascinating, so promising – and so frustrating. It is frustrating at this time because our entire society and its laws and games are based upon the very opposite assumption: that all men are created equal. The great man who first uttered these words hardly meant them in the sense our society is accepting and applying them today. Education, medicine, diets, exercise, rules for behavior, even clothing – all are planned, designed, built or manufactured for this mysterious dragon that never existed: the average man (or woman). Oh, how could any woman be average?

Just as our environment treats every man and woman differently so are they different with different tastes, different needs, different demands and different contributions. Yet, nearly every one among us marches around with a firm concept in his mind of just how any man or woman should be, what he or she should eat, how to behave, how he should belch, dress himself (or herself), think, feel and act. And woe be him or her who falls short or long of the neighbors' standards and demands.

One who taught me deeply about this was the aging King of Afghanistan. In 1928 I was staying at Ankara Palace, a superb – and only – inn in the just-then-emerging capital of the "Young Turks". One morning as I was relaxing in this unobtrusive elegance, the hotel manager approached me, bowing ingratiatingly: His Majesty the King of Afghanistan would arrive that afternoon and, with his retinue, occupy the entire inn. Would I please, kindly, find other quarters.

My shock was as if the Empire State building in New York had suddenly blown down with all the chunks and pieces falling on my head. How did one go about finding "quarters" in this less than half-finished, emerging capital?

In a daze I gathered my things and took off in a taxi.

It was late in the afternoon when I discovered that my only pair of decent shoes had been left at the hotel. I returned, knocked at the door of my old room and – since there was no answer – entered. The room was half dark and seemed to be empty. I headed straight for my bed, bowed down and peeked under it.

I sensed more than saw that I had company. I turned and saw an extremely dignified elderly gentleman on his knees, helping me to peek. As we both found nothing there, we looked at each other quizzically. I hastily explained that I had occupied this room until this morning, had been told to leave because the King of Afghanistan was expected and in a hurry had forgotten my best shoes.

The gentleman nodded, with a troubled look. "You know, " he said, "that was not very nice to ask you to leave just because a king was coming. Kings ought to be told more frequently what inconvenience they cause." The gentleman helped me search the room, stood on tip-toe looking on top of shelves, came down on his knees peeking under chairs. Finally he looked searchingly at my feet.

"You know," he said, "it just seems to me you must use the same size of shoes as myself and it just so happens that I have more of them than I really care for and can use at the moment…"

He opened a monogrammed leather suitcase and there were six pairs of various types, neatly stacked inside. "Do you think any of these will do?"

"No, no, thank you; I wouldn't dream of …."

But he had already taken out a luxurious pair, set me down in a chair and tried them on me. From his kneeling position he smiled up at me, "A perfect fit."

At that moment there was a perfunctory knock at the door and a splendidly uniformed dignitary entered. He eyed the kneeling figure at my side and exploded, "But your maj–ma-majesty!"

I walked out from that hotel on air, in a King's shoes, recalling that old Indian saying, "Judge no man until you have walked a mile in his moccasins."

THE SWALLOWED ENVIRONMENT

The environment we eat seems to worry some people more than what we breathe. Yet, human adaptation is so wide-ranging that if I should frankly tell how little food some persons have taken in, and lived on, and how much has been consumed by others, of what narrow variety, I should be called a liar. I can't afford that this early but at the end of the chapter I'll have done it.

It seems that our pampered, over-protected and overfed generation has lost some of this adaptiveness. In Wyndham, in hot northern Australia, for example, where no vegetables grew during my visit in 1923 and the only food was beef, many developed what they called "dry-rot" in the bush. Some died. Even in more civilized areas the deterioration of soil and the race to harvest as bulky products as possible as early as possible have reduced food values and added hazards. This is countered by a surge in vitamin and mineral intake and this was also why the French Government instigated research on why their bread had become less satisfying. Luckily there is a lot of rethinking done in growing and manufacturing of food and, as with all such things – by private entrepreneurs harassed and often persecuted by government agencies dominated by archangels of the hierarchies.

Will we succeed in spite of the pressures of hierarchies? Or must these be removed or remade? Give a good man the authority to act and in two years he will have them remade.

Not one of the "diets" offered on the market suit more than a few people for a short time; some suit nobody none of the time. But the inventors of the diets have to sell books so they generate the conviction that their prescriptions fill all.

Both the diet fans and their detractors will be equally flustered by what the prestigious Centre International de Recherche Biologique did in Africa. A group of workers deep in the Sahara Desert were so perfectly isolated that their intake and output could be accurately measured. The scientists measured the contents of magnesium in the workers' bodies at various times. All human bodies contain small amounts of magnesium. From analyzing the outputs it was found that tiny portions of magnesium left the bodies. No magnesium was taken in during the time considered. But all the workers still had either the original amount of magnesium at the termination of the experiments, or more. The scientists found no other explanation than that magnesium had been produced from other elements, in the "factories" of their bodies. This was in the early sixties. Later studies appear to have confirmed this. Or, if the reader has some other explanation, please speak up and don't be coy or impressed by the great name of Jacques Ménétrier , the founder and responsible mind behind the Centre.

To some, transformation of one or more elements into another element still seems so far out, so impossible, that they can't accept it so since we have already embarked on the way to cuckoo land, let us take a further step and this time not entirely with the support of Jacques Ménétrier : Sometimes some people by a mere thought seem able to produce even more incredible feats. There was in Europe some years ago much talk about a Theresa Neumann who for many years did not eat at all yet remained in good health. She was a Catholic and the Catholic church assumed the credit and helped publicize Theresa's feat. In India today two women, one in the seventies, the other over a hundred, are reported not to have eaten anything since they were in their twenties.

Some years ago the Medical Association of India was informed and promptly looked into it to find the leak and disapprove an improper claim. The Medical Association is still at it, trying to find the leak. It is almost impossible to find anyone now who will even admit that this case exists and is being investigated. If no explanation comes forth from the Medical Association, the optimists who want to believe that one can live not on bread alone, will conclude that the old girls really did not eat. The pessimists who cling to bread will deem that the matter was so murky the Medical Association did not even want to talk about it. So we shall never know.

What are the girls' own stories?

I have heard only one. The hundred-year-old tells it herself: "I was a fat slob in my twenties, eating and eating and unable to reign in my appetite. I prayed, in tears, that I would be helped. One day a strange man with luminous eyes crossed my path and I felt he wanted to tell me something, so I stopped. "Your prayers have been heard," he said. He gave me some exercises, told me I would eat less and less from then on and advised me not to worry about a thing. I followed his exercises and everything turned out as he had predicted. After some eight months I stopped eating altogether and haven't touched food since."

The man she met, she indicated, was a yogi. What is a yogi? A man standing on his head? A woman (yogini) in a lotus position? That is the popular concept. A yogi is one who practices yoga and yoga has about the same weight and meaning in India as "science" in the west, except that yoga means more to most Hindus than science means to a westerner. Yoga is a name covering all aspects of science and all approaches to science from time immemorial. Many yogis would say that all religions sprang from yoga (though they no longer truly represent it) and all sciences began and are still represented in yoga.

A close friend of mine whose father was a Hindu student of yoga and whose mother was American hiked deep into the Himalayas – fired by his father's tales about the region – and found a man who quickly disappeared into a cave when my friend approached. My friend waited patiently outside the cave. Finally the man came out and asked sternly, "Why did you come?"

"To find a live teacher."

"Why do you seek a teacher when you already have one – your father?"

"My father has passed away. I just wanted to see, again, a real live one."

"Come in then," said the man and invited him into the cave. They had a long conversation during which my friend asked, "I see no food around here. Do you ever eat?"

"Yes, I sometimes eat."

"May I ask, when did you eat last?"

"At the last Kumbha Mela."

The "Kumbha Mela" is a sort of super fair held in one of three Indian cities every twelve years. Many saints and hermits come down from their mountain caves or hermitages at the time to meet the people who arrive in millions. The last Kumbha Mela had been three years ago when this conversation took place.

I brought another friend into the Himalayas in 1959. I hoped to see if a yogi could straighten out her mental difficulties. In the view of the Hindu I was too eager: I started to look for my yogi even in Mussorree, a resort town in the lower Himalayas with easy access except that, perched on the verge of eternity all along the approach road you wonder how far down through the air you'd stay alive if the vehicle slid on the slippery rock and dived.

I hiked, enthralled, through the fern-covered roads and came upon a large campsite with a police guard.

"You want to see the Dalai Lama?"

"Not especially, except if he craves to see me. Actually I am looking for a yogi."

"A yogi?" The policeman smiled benignly. "This is a resort town. There are no yogis here. But you may inquire at the Government's Tourist Bureau. They'll tell you where you must go if you hope to see a yogi."

The man at the Government Tourist Bureau looked up at me, "It is strange you ask. There was a man in here this afternoon. I don't know, of course, but I am wondering…"

"When was he here?"

"At four."

"So if I come at four tomorrow, he'll be here?"

The man laughed, "Well, why not try?"

A man in a long white robe and a crew cut was sitting on the sofa reading a paper when I arrived at four the next afternoon. The

Government man behind the desk greeted me with frantic signs. I sat down beside the man on the sofa. He continued reading his paper. As no opening offered itself, I turned to him and blurted out, "Pardon me, sir; are you a yogi?"

He turned and smiled at me, "A yogi means one who tries to understand life, doesn't it? So we are all yogis, aren't we?"

"Yes, well, I have a definite reason to ask you. I have a friend staying at the inn here, who is not quite well. I wonder if you would kindly come and see her."

He leaned back and closed his eyes. Had I put him to sleep by my boring request? But after a while he opened his eyes again and they looked sad, and his face was drawn in sorrow.

"I cannot go and see her for, you see, she does not want to see me."

(I now remembered that she had said before I left her, "Don't bring any of your so-called yogis here or I shall throw him out!")

"But," he continued, "I shall give you two exercises that would help her, and you can give them to her better than I can. You may think she cares nothing for you, but she trusts you more than anyone else."

One of the exercises was an easy relaxing one, and he explained that relaxation was more than most people understood. By relaxing bodily you released your energy for healing inner, mental or emotional matters. The second exercise was a neck-breaking job, so I said, "But this is impossible. I could never do it."

"No," he replied, "You can't but she can." He then explained how this straining exercise would press appropriate substances out of certain glands into the blood stream and flood and cure the defective parts of the brain.

"Your western science," he said, "operates mainly with drugs supplied from the outside. In your own good time you may discover and appreciate the great storehouse of agents inside your own bodies."

"Will she actually do these exercises?" I asked.

He sank back and closed his eyes again. When he finally came back he looked sadder than that first time, his face painfully drawn. "Three times," he whispered, "and that will be all."

I fumbled with some notes, wondering how much would be just right. He smiled.

"Keep those, " he said, "You may need them where you go. I cannot use them where I go."

"Will I see you again?"

"We are together always, didn't you know?"

Why are so many yogis living in the wastelands of the Himalayas? Why man, the ecology! Fresh, crystal-clear air, cool streams of benign, invigorating water, all far removed from the hustle and bustle of men and women overproducing, overeating, over-discarding.

The visitor on foot, however, will regretfully notice that before he enters the sacred region of hermit yogis, he has to pass the approaches of the pilgrim crowds. In Bela Kushi, for example, where the foaming river of the Aknanda River may tempt the wanderer to bask in its lively liquid, he finds, on the flat rocks along the edges, wastes that in America are confined to the sewer system. Thus every heaven has its unattractive surroundings or outer courts where the seeker must dwell for shorter or longer periods before entering the gates of bliss.

Is it possible, then, that the solution to the ecology riddle – or part of it at least – is man himself? Perhaps with the addition of a few species of animals and plants? That they should limit eating and playing to what the yogis do, meaning to provide the greatest health and perceptiveness? So that this would not only aid ecology but at the same time make life more interesting and provide us with more options? Has a proper balance ever existed? Possibly with a few, but hardly with the many. Our present crude predicament has brought us one profit. We have been forced to look and think. We have been induced to look beyond our narrow borders, even beyond our narrow century.

A thousand solutions are possible and obtainable. We don't need to narrow down our choice yet. Let's just move a bit.

THE PRINTED, THE VIDEOED AND AUDIOED

O ur overwhelming information and entertainment environment can be fully appreciated only when you miss it. I was
in the northwestern Australian desert drilling for water, a
handyman for Charlie, the well-digger. Our only company were a
few kangaroos and a lost lamb now and then. And just at this time
when civilization, libraries and radio were so far away I doubted I
would ever see them again, and TV was hardly invented yet, just at
this time I developed a terrible longing for some word about meditation. Sleeping under open sky, on the desert, and surrounded by
sand all day, with a few spinnefex tufts for decoration, an old longing
exploded into a fierce passion. Like so many a timid youth, I thought
I needed instruction.

Printed words.

I was sitting in almost-lotus position after work one day. I was
radiating out in ever bigger circles this fervent and, I thought, hope-
less desire for relevant, printed words. I noticed a leafy object bob-
bing and sailing across the sand. It came closer. No, it was moving
away. No, it moved closer again. I waited, watching it with detach-
ment, weighing unemotionally in my mind whether Fate might be
involved. Finally it jumped right into my lap.

It was the printed word, all right, the central part of a page of a
book worn to almost a circle around the edges from hopping and
bopping against the desert sand – for how many miles?

I began reading. Within that circle was a succinct and complete
explanation of how to develop your mind and tame your emotions
through meditation. I felt good and a bit proud and as if in mischie-

vous response the desert sand seemed to rise and break and a head popped up. There was a frown and a stare and an impatient voice. "Don't you know that all you need to know is right there inside yourself? Don't you think I have better things to do than hustle bits of paper across the sand?"

Charlie, my dumb boss, hadn't seen or heard a thing.

A psychiatrist gave me a new word: Hallucination. I had heard that word before but now I had to pinpoint it, define it; to correlate it with that head in the sand. From the environment of prints and learnedness came this word to join and aid the environment of sand and heads: Experience. So now, in addition to my confusing experience about heads in sand, I had a no-less-confusing word that nobody could define. They could only give me more evasive words that confused me still more. The environment of print had doubled the confusion left by the environment of experience.

Does print ever enlighten? Less than usually believed – and only when the mind has worked up a receptacle for some specific kind of information, and only as long as this specific bit is transmitted – and nothing else.

I absorbed print in one continuous stretch of fourteen years from first grade until I was in the middle of an engineering course. Then, sick to the bones of print, I had an eighteen months' break in military service.

Afterwards I continued the course, refreshed but still sick. I finished with better grades than I deserved through taking three weeks off for swimming and fooling around just before finals, instead of reading till the last minute. Those who did had a high percentage of ulcers, nervous breakdowns and flunkings.

A better way than a continuous many-year stretch of print would be short courses, distributed along a person's career as and when relevant to his work, as the Armed Services give as well as many industrial companies. Even this crams too much print into too-long sessions. The healthiest learning process for most people is the old apprentice way. You learn as and when you work. There are a few, very few, who by character and talent are equipped to take longer sessions, even continuous schooling up to a degree. But the present pressure for degrees among students and their parents as well as among employers is generally unhealthy and inefficient and particularly doesn't provide the ecologists we need.

Beside the hazard of absorbing too much print at a stretch there is the question of the teacher. Who is fit to cover any subject fully in our complex world? Who, in addition, can communicate with, teach, entertain and excite every student coming to his class? Any lecturer, even the most eulogized, suits only a few. What about television? A little more show and the better lecturers can make the tapes, though basically it is the same: Cramming concepts and word pictures into the student's ears instead of letting him do it himself with hands or mind.

Those who know little about the modern working place tell us our present complex civilization requires long continuous schooling. The opposite may be closer to the truth. Our technology changes so rapidly today that any lengthy course may be obsolete before it has been completed. Besides, new aspects of explaining, naming, presenting the new findings can best be achieved (sometimes only achieved) by the specialists and experts working with or manufacturing these new things. Tenured professors could not possibly keep up with the torrential developments.

A new and expanding teaching instrument is the seminar. Teachers and students talk, discuss, bandy about ideas and concepts at informal sessions. This is different from the old lecture systems and certainly better suited to our present technology.

At the Beijing University the seminar system seems now to be the only method used and has replaced lectures altogether. Another characteristic of the Beijing University is that students are only admitted after a successful period of work. Those who won't work, or haven't learned to work, are not considered fit to study and become leaders in the view of the present Chinese. They also reason that one who hasn't learned something about the society in which he lives and by the grace and work of which he can eat and function, one who does not know all this from experience, has no business at a costly institution of learning.

To satisfy the immediate needs in present-day China, the Beijing

University admits for every ten openings four farm workers, four industrial workers, one commercial worker and one soldier. Other nations would of course choose differently. Many insist that some subjects, at least, are basic for all sciences and can and should be studied continuously; for example, mathematics which is "always the same".

Mathematics is still a desirable subject and in some cases a necessary tool but it is certainly not "always the same". It changes continuously, sometimes for the better; often for worse and has never been properly presented. The minute it will be, not one percent but thirty per cent of us will enjoy, understand and use it.

From modern scientific journals it would seem that mathematics is quite widely used today even in social sciences. This may be an illusion. It has become fashionable for every scientific contribution to be heavily peppered with math. The many scientists who have little mathematic talent hire mathematicians who – on their side – know very little about the problem discussed. You are reading between two unrelated poles, the specialist's and the mathematician's. The few scientists who do master mathematics may come a bit closer though usually they have had to water down their subjects and sacrifice accuracy just to swathe their themes in math robes.

In a sales brochure a new gadget may be so well explained that not only the unit itself is understood but also the environment in which it functions, its reason for being. Do you need a degree to understand these sales pitches? People without any degrees, any formal schooling are today handling the most sophisticated equipment, testing, designing, inventing, creating. They understand the printed word without having wallowed in it.

We are talking about the kind of brochure mailed to a customer upon request, not the ads arriving, perhaps eight at a time, through eight different mailing lists and which is termed junk mail. If this practice be persisted in, our postal system will break down. And for what? The junk mail ads are rarely informative, mostly uninspiring, sometimes bordering on the moronic. They make you doubt that the inventor of print had a good thing.

Only for a moment, of course.

I live in a happy world of a thousand books selected by myself, crowding bookcases, spread over chairs and the floor. My only regret is that I haven't time to read them through. My publisher friend has other regrets. So many excellent things come across his desk from the millions of known and unknown writers. To remain in peace with his shareholders he has to submit to the judgment of his Sales Department which has to reject great, evocative offerings that could have entertained, amused, charmed and lifted the nation. So, my thousand books do not represent the whole effort – may not even represent the best.

Topping this, the Chinese have a saying that is increasingly appreciated as one's years pile up: "True wisdom cannot be found in any book."

In regard to the printed word, the only relatively happy people seem to be either the publishers and editors who read and enjoy everything – whether published or unpublished – or those who read nothing. Neither is caught in the trap of the printed word.

The printed environment and the videoed and audioed, are similar to the air and water environment in that basically it is good and almost necessary though certain parts are poisonous, nauseating, make the eyes smart and the nose run. We hope to reduce the foul part so the healthy part can live and breathe and express itself.

As far as air and water are concerned, we were just beginning to see that the pollutants are results of our own actions and behavior and expressing ourselves. The printed environment is more obviously our own creation and may help us see that the environment and man may be considered one and the same complex multi-being. The environment changes with us. It becomes better and healthier when we become better and healthier. The encouraging thing is that we are becoming healthier, more conscious. We are working ourselves out of the doldrums. The foul air and water and the foul print is helping us see ourselves and helping us change. We are growing up.

Some books and magazines are ably reflecting this new, warm concerned in-feeling with man, his society, environment, and his yearning. They demonstrate the newer, closer, warmer touch created, in part, by our rude awakening amid deadly pollutants. These books and magazines, in turn, influence newspapers, TV, radio and plays, individual attitudes, the general atmosphere.

Prints, videos, audios, are part of our neighbors and not merely close neighbors. They may be from thousands of miles away, from minds we have never seen. There is one advantage with this kind of neighbor; if you don't like the book you don't need to read; if you dislike the TV piece you can turn it off. After having been exposed from babyhood on, however, how do you know what to read, what to turn on and what to avoid? That is the trouble with the print and video environment. Another drawback is that you cannot talk back, not unless you have the facilities for launching a book or a set of articles or a TV talk or show of your own. Therefore we have been blessed with direct neighbors, humans in the flesh who come knocking at your door to borrow your baseball bat, lawnmower and camper and who may ask you for a cash loan in addition to make sure you'll not forget. This all guarantees a two-way flow! Action and reaction; one word leading to another; peace and war.

NEIGHBOR

In Fremantle in Western Australia I had been adopted, for a while, by a Norwegian family. One evening they introduced me to a formidable Swede. Swedes, being usually more affluent than Norwegians and occasionally more generous, my new acquaintance soon came to the point. Taking me into a room where we were alone, he told me that he knew I had come to Australia with practically no money, that I had worked on a farm for ten shillings a week, so he wanted to offer me a loan.

I was embarrassed. I did not see how I could accept a loan since I had no collateral and no firm income in sight. Now he became eloquent, expansive: He trusted me completely, he said, I was the type who would soon get regular work and then I could repay him. It would be a pleasure for him to give me a loan. I just had to say how much.

I stood firm: No collateral, no loan. We parted, praising each other ever after and never meeting again. Both were embarrassed. Good neighbors.

The neighbor environment may be further pin-pointed by relating what went on before and after that meeting with the Swede. My well-paid engineering job in Borneo, Dutch-owned a that time, was too sedate for my ebullient young body, so I quit and boarded a Chinese tramp, Ban Ho Guan, in Bandjermasin (meaning "too much water".) Its skipper, a bearded Frenchman with an iron hook for a right hand, described to me the battle of Strassburg, where he had lost his hand. He was so absorbed, he ran the steamer right into the side of the Barito River. A treehouse, with an angry Malay woman, nine jubilant kids, a black grunting tamed boar with eight young, came plunging down on the foredeck. The skipper cut short his eloquence, locked himself into his cabin and I never saw him again. I spent the days with the Chinese crew who made tea in diminutive earthen vessels, grinning amicably and talking Chinese to me without being in the least disturbed that I did not understand a word, but responded with some English of my own, to which they nodded eagerly. They never called me an Amk or a Nork, possibly because I never called them chinks.

For five days they were my only neighbors. I never had any better ones.

In Singapore the British authorities seemed not yet to have learned that spies always travel first class, for they showed extraordinary interest in this person arriving by a Chinese tramp and took me ashore in their brass-decorated launch and further on in a shiny government Rolls Royce. The citizens of Singapore, just then expecting the visit of His Royal Highness the Prince of Wales – who later was to become the Duke of Windsor – watched this humble facsimile which to them looked like the real thing and roared their welcome. The official at my side lost his cigar from the shock but recovered it at waist height. He was a good neighbor.

In Singapore I boarded a cattle ship bound for Wyndham and later Fremantle in Australia. We loaded the cattle in Wyndham, this hot North Australian town where people eating beef exclusively died from "dry-rot". The cattle were pushed aboard by electric prodders and many died during the journey to Fremantle.

Before going ashore I faced my cabin steward., "I have one pound and five shillings left. That is all I have in the world. So I am sorry I have only one pound to give you. Five shillings I figure I need on shore."

I hoped he would take only five shillings and leave me the pound. He said, "That is quite all right, boy. A pound will do." So he took my pound and I landed with five shillings. There was thrill to this. How would the Australian environment and neighbors take care of me? I went straight to an employment bureau. They had a job for a farm boy "one hour's train ride from Fremantle." It would be twenty shillings a week, said the man, and I had to pay him ten. I gave him my five shillings, promised to pay the rest later, found the railroad track and started walking.

Early in the morning I reached the farm. Saunders was just having breakfast. He was about to go to work for the railway since the farm couldn't support him and his family. He decided to stay home that day to instruct the new boy. I told him the employment bureau had told me I would earn twenty shillings a week.

"Those cheaters," he exploded, "They knew I could not pay more than ten. And they took ten shillings from you?"

"They tried," I said, "But I had only five; promised to pay the rest later."

"Promised my eye! You paid them just right."

There was wood to be chopped. I had been good at that back in my native Norway. This was different: Ironwood it is called, hard as iron. Saunders had gone back to the house. His nine-year-old daughter was watching me. Suddenly she cried out, "Daddy, daddy, come and see the new Jack – he has turned green."

Papa Saunders came running and looked me over sternly as I leaned against the barn door, ready to faint. "Jack, I didn't know you were a drinking man!"

But he decided to keep me on when he had heard my story; no dinner, no breakfast and a walk along railroad tracks during night. After a few days he even gave me a nod, "You don't seem to know much about farming, Jack, but I must say this for you: you do know how to handle a horse. Ours is a mean one. You handle him like a pro."

I knew what it was. Having never handled a horse in my life, I approached this one humbly. He felt my respect for him and liked me. We were also good gold-brickers, both of us.

The working day was from four in the morning till eleven at night and never any time or place for a bath. This worried me, so once we were working on the well, I simply "fell" in and squashed around a bit. Saunders was calling soberly from the rim, "Your sprawling around down there doesn't exactly improve our drinking water you know."

The event may have sobered him. That evening after having exchanged glances with his pretty wife, he said, "Hm, Jack, you wouldn't run away with our silver, would you?"

I laughed; he continued. "You see, our former boy did and that is why we had to let you sleep in the fertilizer shed." (and his calves licked my toes, sticking out of the shed, and so awakened me at four every morning.) " But my wife and I have decided we'll trust you to sleep on our porch."

Fringe benefits. And at the next meal after Saunders had given me one slice of bread, as usual, and seen it disappear down my throat, he asked, "Jack, you want another slice – NO?!" And he began to ask me in advance whether I wanted butter or jam on my slices.

And one day he explained it all by telling me the name of his farm. It was a Latin name and it meant "I Strive Hard".

Saunders had been a bookkeeper at the railroads. He became tired of sitting behind a desk just as I had in Borneo. He had bought this "I Strive Hard". And he did.

But to spend the rest of my life at 10 shilling or $1.40 a week with only one slice of bread per meal with butter or jam, and a bath only when you fell into the well, didn't seem necessary. I was gentle when I left, said I would like to go down to Fremantle and tell those so-and-sos in the employment bureau a thing or two and perhaps I would be back, for I liked them both.

I walked right down to the harbor in Fremantle, sat down on the tip of a pier and splashed my feet in the cool salt water. I intended to drop right in, clothes and all, and swim right to India or somewhere when I heard steps behind me. Without turning I listened to those steps. I figured from the firm setting of the feet on the ground that here was a man of the sea who came on a mission of mercy. The steps stopped. There was a short silence, then, "I want to talk to you."

I got up and turned around. I had been right. This was a man of the sea. His face revealed his thinking as clearly as an open book. Old sailors did not learn to swim. It would prolong the agony if they fell overboard far out to sea. When they see someone sitting on a pier, his feet playing in the water, they have only one thought: this man is starving. He is desperate. He is planning to drown himself. Suicide.

He sank his blue steel eyes into mine. He reached out a hand; there was a pound note in it. "Go get yourself some work clothes. We are sailing in an hour."

"Sa…sailing"

His red-haired fist pointed to a schooner at the next pier, "Thar she is. We go north to the wool stations."

He turned and left. This was how he got his crews. And saved them from drowning.

There was a Scotchman and a Swede on the schooner beside Skipper Melsom and myself. A close neighborship. No time for nonsense though lots of jokes, an artful system of positive reinforcers in operand conditioning a la B.F. Skinner. Otherwise we would have killed each other. So B.F. isn't new but he's working.

The Onslow River has deposited silt miles and miles out into the ocean. We anchored outside that silt and Skipper Melsom chose me to row him to shore, which caused him to give me his first compliment, "Jack, there isn't much to say about you, so let me say it now: You do not know how to row."

Ed Patterson, the gentleman of the Northwest, was waiting for us with a pilot. Ed owned one of the medium-sized stations with 400,000 sheep. The biggest had a million. We rowed the pilot out to the schooner and he took it in, zigzagging among the sand bars.

Skipper Melsom, without asking me, offered me to Patterson as a jackaroo, lowest paid hand on a sheep station, adding that I could row. "There isn't much rowing to do around here," replied Patterson, "besides, we are overstaffed."

"Overstaffed? Now, in the shearing season?"

"Yas Melsom. Overstaffed."

Patterson was in a hurry. He wasn't going to let the speed of unloading be decided by a city-slicked crew. First there were salt bags to be unloaded. He placed himself at one end and looked around, sternly. Johansson, the Swede, lumbered up to the other end. Patterson, setting the pace, the bags more flew than moved up from the hold and on to the wharf. After forty minutes Johansson sat down, panting heavily. McMurdo, the Scot, took his place. He was breathing hard when the lunch bell saved him.

After lunch there was no one but Skipper Melsom and myself available. I couldn't risk the skipper having a heart attack. Wasn't it just my luck to be matched against this superman when he was rested, after lunch?

After only 15 minutes I tasted blood in my throat. But I had made up my mind: I was not going to give in to this blustering bull-gentleman of the Northwest. Soon I knew neither time nor pain. I was going to stand there throwing salt bags till the end of the world.

Patterson sat down, wiped his brow and panted, "Jack, you're hired."

He drove me through his desert that night, no roads; sometimes not even a track and kangaroos dancing in the headlights. Three pound ten a week, and all the mutton you could eat for thundering along the desert on horseback, donkey or camel, chasing sheep or setting up fences; digging post holes with a soup can or with two at once when you became an expert and setting the poles by shaking the sand around them with an iron bar until they stood like towers of strength and durability. Watching the donkeys, twelve of them, sing their serenades each midnight before manager Red Tito's bungalow until he rushed out, stick in hand and his old-fashioned night shirt swirling around him. This was when the twelve took off with their most respectless, gay hee-haw-hee haws. Since I was the only one who didn't laugh at this spectacle, the manager began talking about grooming me as his successor. This was the signal for me to take off and join Jim Whitaker on the mail team, and when I had taken apart his Ford engine and couldn't put it together again, I quickly switched to Charlie, the well-digger. The two of us plus two horses operated in the remotest parts of the bush or desert.

I ran barefoot but Charlie wore socks and boots and this saved us from a wild camel who came rushing into our camp one evening during the mating time when he-camels ungallantly smash anything they see with their big hooves. This time the camel's olfactory talents were so overwhelmed by the smell from Charlie's feet the poor animal turned and galloped off before it could hit us.

Worse was to come. There were again, that night, muffled drum beats of hoofs in the sand. It was not a camel this time; it was a shining white horse bursting into our camp like a vision from Heaven, for, on its back – a woman amazon such as you haven't envisioned in your wildest dreams! Features hewn in marble, curves knocking you over, making you seethe and boil; hazel eyes fastened lingeringly on your fringed pants and stubbled chin.

Senseless, I ran after her, threw myself down in the sand and kissed her horse's hoof prints. "What are you doing!" bellowed Charlie. "Kissing her footprints," I hollered back. "Her footprints, my eye! You're burying your face in a damn dirty nag's hoofprints – that's what!" So I was.

It was on my return to Fremantle, staying with my old boss Skipper Melsom, that the affluent Swede entered my life and made me embarrassed. Embarrassed and reflective. I looked over my situation and found that I really needed a loan though not from a Swede (my being a Norwegian) and not from one expecting me to make good in Australia. I wrote to Benard Chabot, my French-Malay friend in Borneo who had literally begged me to accept a loan from him. I didn't accept then, but now I wrote him humbly for twelve hundred dollars for traveling back to my home, Norway – at 8% interest which was high at that time. He immediately sent me two thousand, saying I had probably not asked for enough. I repaid him in the course of three years. He wrote and thanked me and asked me to please visit him some day. He had had many people returning part of loans he had provided but never had anyone paid it all back and never, never, never, never with interest. I know that others have done that too, especially to banks, which become unpleasant if they don't get it all back with interest. If more did in private transactions too, we would be more like B.F. Skinner: positive reinforcers.

If an excuse is expected of me for devoting an entire chapter to these Borneo-Australian neighbors, it is that both previous and coming chapters may benefit from these rapidly-changing neighbors, and their implications on the closer neighbors: our bodies, minds and souls.

A Closer Neighbor: Your Body

When my body felt so uncomfortable sitting at a desk in the Borneo heat, I exchanged my eight hundred dollars a month job for a $1.40 a week splash in Australia. My fellow men will laugh and ask if I didn't know of tennis? Swimming? Calisthenics? I did play tennis twice a week and a crowd of native spectators roared every time I managed to ht the ball. I did swim in the Amandit River and got dysentery. Apart from that all of my fellow men who swim and play golf and tennis are old and used up at the age when they decide things in our society, such as ecology, pollution control, law and order.

They decide with a yawn and while tired, waiting for an early death, like the old Professor of Meteorology at the University of California. He had played tennis too. Or the once-upon-a-time head of our Federal Reserve System that controls our money and does more than any other institution in deciding how many shall be permitted to work, what scientific research we are permitted to begin or continue; what armaments we shall be allowed, to secure our independence and leadership. Our most honored teacher of Economics, Paul A. Samuelson, the Nobel Prize winner, was once asked what he thought of this head of the Federal Reserve System and his management of our money. Paul thought a while, then said, "Well, a man who is so fond of tennis cannot be all bad." Indicating that tennis is not enough to make a man sound of judgment.

What is? Has the body or its exercises or even diet anything to do with it? We know of many people of extraordinary judgment who suffered serious diseases or had no exercise. We have to admit, therefore, that there are no general rules. The truism of individual differences still stands.

Generally, though, remarkable success is just now being achieved in restoring body functions that had been reduced and sometimes almost strangled by what we have called "civilized" living. In the bargain it was found that body and mind have an almost incalculable influence upon one another. Particularly successful in this quest have been medics, biologists and psychologists who drank deeply of the treasures and traditions of the past, in India, China and the Middle East.

Just at this time the ancient Chinese science-art of acupuncture, briefly touched in a previous chapter, has entered American hospitals in New York, Michigan and California. Dr. E. Grey Diamond of the University of Missouri wrote about it in the *Journal of the American Medical Association* in December 1971, after he and Dr. Paul Dudley White, the cardiologist, had visited China together. Dr. Diamond dwelt particularly on the anesthesia effect, a rather recent development in Chinese acupuncture and perhaps not the most important aspect. Major General Walter R. Tkach, President Nixon's physician who accompanied the President to China, also stressed the use in anesthesia. Of course, to be able to perform major surgery without the risks and side effects of our usual pain-killing agents is an achievement in itself. Our physicians may wish to introduce this one aspect before going on to more general practice of acupuncture, which has been known to improve or cure quite a number of serious diseases.

If we ever do go the whole way, a distinction will be felt between the ancient Chinese concept of health and disease and our concept. The real acupuncturists disdain our list of fearsome diseases, and see, instead, simply an imbalance of life forces coursing through our system. In this way they are in accord with the yogis of India and the Sufis of the Middle East. The forces that have to be balanced have no corresponding place in western medicine. Neither have the channels or leads through which these forces run. These channels are pinpointed by the acupuncturists in great detail. Their systems are no less complex than the systems of western medics but the complexity is in an entirely different area. While the Chinese accord ample room for our western science and systems, gradually integrating them, our western concept has no room for the Chinese (and Hindu and Sufi) systems yet.

Acupuncture is based on concepts that are not included in conventional western science and cannot be "integrated" in our science as it now stands, except that the Humanist Psychologists (comprising mainly forward-looking M.D.s, biologists, archeologists and other professionals) and related groups have during their experimental work stumbled into some of the concepts and facts underlying acupuncture.

These concepts include forces or energy bundles that have not been registered by any known instrument. Some people appear able to see or feel these forces and perceive their channels. Most acupuncturists don't. They have simply learned their profession, carried on from generation to generation. Those who can see or feel these forces to any useful extent have always been few and it is doubtful this faculty can be learned. It seems like the talent of music, except that it is more rare. The Menninger Institution recently acquired the services of an accomplished yogi, Swami Rama, who not only has this "feel" but is able to demonstrate it in terms western science, or at least some of our scientists, can understand and appreciate.

The emerging recognition is that our knowledge in medicine, biology, human and animal anatomy, is so sketchy that we may say we have hardly scratched the surface. The details of our visible nervous system may have been mapped more accurately than ever before in our modern science but vaster areas and relationships have been unknown to us and possibly only faintly understood and differently understood by Chinese, Hindus and Middle Easterners. To combine all this into a grand total may be the most rewarding project now, possibly bringing shattering revelations.

A human being, or any being, according to the classic Chinese, is not a stable smooth-running engine, is not meant to be, but an extrusion of Mother Earth itself, liable to the same opposite forces, Yin and Yang. All the needles do, when inserted, is to return the body to a proper balancing of the forces when this balance has been disturbed. Disease, any disease, is just that: Dis – ease, lack of ease.

Are we any closer to understanding sickness and health then? Not exactly understanding, perhaps, not today at least. None of the Chinese medics our American doctors talked to said they understood. But they have shown one method that works, to a remarkable extent, on an increasing number of complaints.

Are we at a stage in civilization today when we can heal, but don't know why? This is what many Chinese physicians would claim. In the West, we have an urge to understand, to explain, and some even claim to understand.

Apart from the needles there are herbs, drugs, surgery, vitamins and minerals, protein and carbohydrates. To what extent could all that be replaced by needles? A physician in our time sees a vast field of research before him.

Western physicians, other than Americans, have studied acupuncture for decades. Jacques Ménétrier , outstanding Parisian physician, physicist, electronics expert, mathematician, artist, and author, whose acquaintance we made in an earlier chapter, bases much of his diagnosis as well as cure on a similar system in which he uses electrodes instead of needles.

Russian acupuncturers also often use electrodes instead of needles. They refer to a Caucasian system of acupuncture in which the points on the body where needles or electrodes are placed vary from the Chinese. Are the Russians nationalizing human bodies? If oriental and occidental bodies differ, where should the border be drawn? What about the Turks, belonging to both Asia and Europe?

Speculating on Dr. Ménétrier 's catalyzers, could their effect be compared to that of the acupuncture needles? Would the latter have a catalytic effect as part of their healing or "balancing" quality? Ah, my Western mind is trying to understand and explain.

Other medics, mostly young and yet unknown, have gone to India instead of China and studied the yogis and rishis and, as we know, yogis have not been reluctant to come to the fat of America.

The yoga culture and its healing methods are wider-ranging and older than generally realized in the West. Most people think Yoga is certain posture and exercises. These are all a tiny part of one single branch: the Hatha yoga. Nevertheless, this is important: without any drug, needle or diet, certain simple postures and exercises may dramatically change bodily health, with beneficial consequences on the mind. Then there are the Gnana yogis, yogis of the mind, who use little if any physical exercise or postures but reorient and revitalize the whole personality through their mind. Which of these two methods or systems is to be used depends on the patient or the pupil. In this, the yogis are ahead of most of us: they recognize that humans are as different as night and day. What is poison to one is cure for another.

A third type of yoga is Bhakti yoga, the yoga of devotion or love. The whole personality, mind and body, are cured by emotional input. Then there are the Karma yogis, the yogis of action. Soldiers, statesmen, business men are often cured or satisfactorily developed merely through their actions. There are the Mantram yogis, who develop through repeating magic words or chants, "mantrams", and all of them pay attention to their breath.

Yoga, therefore, is not a narrow special practice or philosophy. It is the whole past civilization or culture of a continent, its science, religion and behaviorism rolled into one.

In the Near East a corresponding wide-ranging discipline or philosophy is and was Sufism, the inspiration behind the Hebrew, Zoroastrian, Christian and Muslim cultures and religions. The present popular fashion of connecting Sufism exclusively with Islam is deplored as much by knowledgeable Muslims as by other scholars.

Sufis have for millennia been living side by side with yogis in India and other parts of the world. Both have benefitted and expanded their concepts. Both, again, are closely related to the Buddhists, particularly to Zen.

A medic who today approaches his task with a broad mind, a minimum of prejudices, faces a whole new world of opportunities. He or she may come to view the entire environment of man, of his body, of his mind. The parts of the body may then look, not so much as separate parts, rather as one coordinated whole. Body and mind may not seem two independent units but one whole. Man and his environment may look as one meshed, inseparable entity.

Apart from the benefits to himself, to many young people it matters quite a lot how a doc looks at body, mind and the environment. Since these young people will inherit the world, and run it, why not pay them heed? Is it any wonder that they doubt the relevance of our authorities, of our education that demands adherence to inherited concepts? Which may turn out to be superstitions? Many of the young may not know any good alternative, but when experienced men do come forward with alternatives, surprising numbers of the young eagerly listen and often follow, as in the case of economist John H.G. Pierson, ecologist Barry Commoner, philosopher Oliver Reiser. These men presented, not so much a new theory, as life the way they saw it; not bits, pieces or dreams but a real surging stream.

When I was young myself I met some such people, a wise old owl-aunt, a Hindu singer who heard the song and longing of every soul he met, and an Egyptian scholar who had been head of all Egypt's libraries and appeared to have read all the books, yet, often quoted a Chinese saying, "True Wisdom cannot be found in any book."

Ali Fauzy Bey summarized some of his vast knowledge in gentle words about the one Near-Eastern religion that was called Hebrew, Zoroastrian, Christian or Muslim depending on where and by whom it was presented. This one religion of the Near-East, he added, was now gradually being merged with the Far-East Hindu-Buddhic faith. In this he was supported by Jawaharal Nehru, who, though not a narrow religionist, must be considered Hindu, and he called Buddha "the Greatest living Hindu…"

Ali Fauzy Bey called himself a sufi and his face grew dark when he talked about Muslim scholars who had tried to make sufism a Muslim creation. Islam, like Christianity and the Hebrew faith were all sufi creations, he said. He was particularly surprised at the Christian scholars who had fallen for the Muslim "canard" instead of going deeply into the matter and so help Arabs and Israelites see that they had one and the same religion and inspiration.

I met Ali Fauzy Bey in Istanbul. Together we went to the Mevlevi meetings, the Howing and Whirling dervishes, among whose members were in older days most of the military and civilian leaders of the sultanate. When the young Turks under Kemal Attaturk threw out the sultan, they also evicted the Mevlevis, the howling and whirling dervishes. But not all members ran. Some stayed on and came together in clandestine gatherings, which had all the charm of forbidden fruits. Ali Fauzy Bey, the great scholar, closed his eyes and rotated his grey head in rhythm with the music. When he had recovered from his ecstasy he outlined for me, calmly, the trends and traits of coming decades: crowding, ecology, changing morals, dirty cities, a rising from the ashes with new, exciting life forms.

THE GOLD MINE BETWEEN OUR EARS

The gold mine between our ears was Walter Moon's term for mind, our very closest neighbor, in the *Niagara Frontier Purchaser*. Mind is the source and substance of our non-physical environment. Is it really limited to that space between our ears? You may think of a place or a person thousands of miles away. Are you not there, then? Even though your body isn't? You are mind, aren't you? When taking into account known cases of instant thought exchange at a distance, called telepathy, doesn't it seem that mind is a wide, far-reaching form of life, uniting us, subtly, yet firmly?

It is optional, or a matter of semantics, whether we want to use the term mind for everything in us other than the body. Some distinguish between mind and emotions though that is difficult for every thought is brimful with emotions. Some talk about spirit, or soul. Though never defined, there do seem to be stirrings in the consciousness not easily compatible with what we call mind. Could mind be considered just a link, a go-between, joining spirit and body?

If so, it is not a straight or simple go-between. It is so complex and tricky that many lose themselves in its labyrinths and concepts. I told my mother when I was eight that I could jump higher than the tallest house. When she challenged me I fumbled a while then got out of it by asking, "Well, how high can a tall house jump?" It took years for me to discover that even grown-ups play such games, in rip-roaring politics, in speculative business; in parties – to gain attention of the girls – and, forgive me, sometimes even in religion.

The games are not always played just to fool the other guy but often to fool the player himself also. When Paul Ehrlich asked Barry Commoner to stop insisting that the latter-day technology was the main cause of pollution and instead to stress population explosion as a cause, the reasons were complex. Paul must have known that Barry would have witnessed against his better judgment if he followed the advice. But he, Paul, had woven himself so tightly into the population conviction he was convinced this great cause warranted hiding or shuffling other facts that appeared less important to him.

This is how politics, religion and business are conducted every day. And it is a question whether pollution or any other cause can be effectively dealt with as long as we deliberately play that kind of game. Barry Commoner, at least, was not willing to compromise with what he saw as facts.

Both parties twisted mathematics to serve their ends – deliberately or unwittingly?

Grades, ranks, titles, hierarchies, pride, humility and judgments are all games of the mind, often enjoyable; sometimes useful; more often useless and devoid of any ultimate reality. The game of politics may save a person from a limited concept – to plunge him down into an equally narrow ditch. Politics may save a nation from excesses – and also prevent a wise counsel from ever being heeded. The mind jumps quickly from a mere word or gesture to irrelevant and often fateful conclusions.

In the environment called religion, people cling doggedly to a concept which they mistakenly call "Faith." They mean "Creed." Faith is a larger thing, the wordless uncluttered drive that makes a tree reach for the sky and makes man and woman face and conquer the incredible, the impossible. Are all men and women endowed with such faith, a faith that enables them to do the impossible?

Perhaps not all, but many, many more than are apparent in our present society. Why? Are they hiding out on purpose? No, it is just that everybody is running around nervously, trying to make some money to live, either by going into business or getting a suitable job. And some of the best among us don't get that job or are fired from the job they have and don't immediately get a new one. Yet, as everybody knows, there are a million urgent things that need doing and there are people and resources to do them.

What is the matter with us? Why haven't we succeeded in this apparently simple task of putting ourselves to work?

Our minds. Not all minds: some know how to do it, but at present the majority are unable to see not only how we could organize ourselves but also who could do the job. While many minds are good as far as they venture to scan, they have grown up with confidence in specialists. The employment situation, they say, is the task of the economists.

Which economists? Anyone with a Ph.D.? We haven't taken time to look at the schools who train the economists and other specialists. Only very few come through that training with their minds clear and alert. As in religion, so in most sciences, rigid concepts are mistaken for facts; theories are presented – and accepted – as knowledge.

The young faithfully follow in their parents' footsteps until one day they discover that they never learned to use their minds nor did their parents. As a consequence, the society in which they are invited to live, work and play does not even satisfy the most basic requirements. So most of the young band together in stark opposition. Only a fraction of them realize that they do not know how to develop the mind either. This fraction joins grown-ups of their choice and try to find out. The latter group of youngsters are the stuff the future is made of. Now in my seventies and eighties I have benefitted enormously from their counsel and their courage.

Since schools and universities tear down as much as they build up and don't really develop the mind, where may we turn? First, we do not reject schooling. We go through it with an open and critical mind looking all the time for other avenues. The best impulses come from men who have gone through the gamut of examinations and degrees, who have survived these aberrations. (Economists like John H.G. Pierson and John Philip Wernette; Sociologists like B.F. Skinner and John Platt; Psychiatrists like Jule Eisenbud and the founders and members of the Humanist Psychologists.)

As we know many students and teachers have delved into yoga and Sufism, ancient traditions for mind training and – we may add – heart training; cultivation of thoughts, emotions and balance. This trend is called quackery by many people though this author who was familiar with what is called mystics from childhood, has retained the deepest respect for the best in these training methods through-out my engineering and scientific education and my toil and fight through sixty-five countries up to my present age.

Yes, it is possible to train your mind, to make it more open to new ideas, to make it more understanding of other people, other races, other nations. When you meet a person and don't first see a cluster of what you call faults, but a great moving and reaching and yearning being – then your mind and heart are beginning to develop. When you understand that by mentioning what you consider a person's fault you nail him to that fault instead of letting him go on, when you clearly see this, you are on your way.

Isn't this what religions teach?

In the distant past when they were born they may have done so. Today the religion first wants to know what you believe; it wants to judge you on a concept – its concept, before it will say anything good about you, or anything at all. The story of the good Samaritan has been forgotten, but not by all.

Much of the religious yearning today finds outlet through psy-chics and fortune tellers. A medium in Oslo predicted the imma-nent death of an old friend. The friend died. Soon afterward that medium carried her diamond. She was arrested, accused of having contributed to the death. When American psychics predict assas-sinations, desperately hoping to prevent them from coming true (physical, character or loss-of-freedom assassination) what happens to the minds of prospective assassins? They are impressed, may act.

We have found – searching medics have found – that minds reach beyond the brain. How far and in what manner is not generally known but psychics who make tragic predictions take tragic chances.

Sophisticated minds, fond of symbols, find in mathematics an outlet that sometimes may serve as a genuine extension of their minds and lead to solutions such as projecting a trip to the moon that would have been unattainable without the mathematical tool. Yet, there are some who arrive quickly and directly at such results with the help of amazingly simplified mathematics and sometimes without any.

On the other hand, mathematics may lead even a good mathematician into a trap, wrong solutions, portrayed touchingly by our great mathematician, Edward Teller, in *Legacy of Hiroshima*. He not only was trapped but so thoroughly that when his colleague and friend pointed out his mistake he first suspected this friend of making up the whole thing, and accused him of willfully trying to wreck the hydrogen bomb project. Edward Teller finally saw his mistake and apologized.

Astrology is a related younger cousin. While most mathematicians would pinch their noses to avoid the smell, some have actually sniffed at the astrological symbols and ideas and come up with the following comments: Present astrological charts are based on different and incompatible bases; do you look at the heavenly bodies or symbols from the earth? Most astrologers do. Or do you look from the sun? This is called heliocentric astrology and the resulting charts are different. Or are you based on constellations? This is called Hindu astrology, though all three forms are practiced in India.

Then about the twelve zodiacs, is there anything basic or structural in these twelve clusters, or are they merely short-cuts of the infinite number of points on the hemisphere? Has this been researched by competents or is it merely an inherited practice, indiscriminately used?

At the beginning of this century an Indian yogi, Sri Yukteswar, wrote in *Holy Science*, that the Hindu astrology as then practiced was thousands of years off in its calculations, due to a mistake that crept into the almanac about 700 BC. One of the consequences, he wrote, was that the generally accepted idea of astrologers that we are now in a dump of a spiritual cycle is all wrong. We are far up and rising. Some have accepted this correction. Others work on the old assumption.

There is no doubt among perceptive people today about the close relationship of everything living, including man, woman, animals, trees and plants, planets, suns and stars, all linked in a complex, enjoyable rhythm that we should try to feel and know by all means at our disposal. In this quest it is well to know the processes of our minds and not fall into traps. Luckily some of the leading astrologers are doing just that. Dane Rudhyar writes in his latest book, "Astrology, as I understand it, has no concern whatsoever with whether a conjunction of planets causes some things to happen to a person or nation."

In the fifteenth century Paracelsus wrote, "Constellations are subordinate to the wise man. They have to follow him, not vice versa. Only a man still on the animalistic level is ruled by the planets." If you have to be an astrologer, the Paracelsus type offers certain advantages. It leaves you room for building character. Inayat Khan, a Hindu musician and sufi, was asked if it wouldn't be a good idea to have one's horoscope read. He replied with a question, "Where are the astrologers?"

The mind is our closest non-physical environment and the source and also the instrument of all ecology. The more free and uncluttered it is, the more successful of ecological drive. For proper functioning, the mind must be challenged, tested, cleaned and oiled all along.

Walter Moon, in his story about the "Gold Mine Between Our Ears" in the *Niagara Frontier Purchaser,* was telling about a specific use of minds called brainstorming. This is a meeting of people selected for shooting out ideas about certain plans or improvements with no inhibitions at first, no hesitation, no second thoughts – and no criticism allowed either. When a batch of ideas have been thus collected, some good, some far-out and impractical, another select group undertakes to evaluate them. It is used in nearly all types of business now and we are thinking of it as a Western invention. Actually it was used in nearly all civilizations and the Hindus call it an aspect of Karma yoga (the yoga of action) with a bit of Gnana yoga (of the mind) and Bhakti yoga (emotions) added. This author has been practicing it for some years in the US Navy. I also know it from yoga lore.

Through such brainstorming the astounding power of uninhibited mind activity is demonstrated. Psychologists are speculating: what is the interaction between various participating minds? Through systematic as well as intuitional search into these matters the most exciting inventions of the future will be made, to the greatest benefit for ecological solutions.

BEYOND MIND?

A single flagstone crosses a stream into Dayak-land, separating it from the world of the Malays. The Malays live on the plains where rivers move slowly and flood the whole countryside in the rainy season. Dayakland rises, fresh and healthy, cool streams chuckle while they swab rocks and boulders. Along the sides of shimmering mountains run the footpaths linking the villages and the mountains change size and contours as you move along and view them from different angles. The Dayaks fiercely defend their lovely land.

Early in this century, when the Dutch held this country, a Dutch "controlleur" walked into Dayakland with an interpreter and twelve armed Malays. He went in to collect taxes. Nobody ever heard from this controlleur again, or from his interpreter, or from the twelve armed Malay. It was generally admitted that it was unwise to bring the Malays in. There had been century-long enmity between Dayaks and Malays. It is assumed that shrunk heads of the entire mission now adorn village chiefs' belts.

In the fall of 1922 another Dutch controlleur entered, again with twelve armed Malays, notwithstanding what had been said before about bringing Malays to Dayakland. This seems to be the way governments work. I was attached to the party, an engineer sent to explore the upper run of the Amandit River for our irrigation work. We were fully aware of what had happened to the previous mission and that it was unwise to bring in Malays but such is the insouciance of youth that we did not for a moment think of danger, at least I did not.

After a week the controlleur returned. We had not yet reached the upper regions of the Amandit River. I proposed to go on alone. The controlleur said he would send along with me one armed Malay. I asked him to reconsider, feeling this arrangement might benefit neither the Malay nor myself. But he insisted.

Four smiling Dayaks with hooked swords escorted me to the next village, I in front with one of them; the Malay behind me with the three others. In a bend in the path I thought I heard a sound like a hiccup behind me. I switched around and saw nothing – no Malay, no Dayaks. But soon the three Dayaks appeared again. I rushed past them looking for the Malay. They came along, laughed and one said, "Malay gone home. He no like Dayaks." He spoke Malay, the only language I knew. One of them handed me the gun the Malay had carried, with a big smile.

There was nothing more I could do. We continued.

When we reached the next village the four Dayaks stopped two hundred yards away and motioned for me to go on alone. I thought they felt they wouldn't be safe in the village.

Distinguished dignitaries of the village seemed to be gathered before the community house on its high stilts when I approached. They looked grim. I stopped ten feet away and looked them over with what I hoped was a gay smile. Was there no opening in any of the faces? Oh yes, a deeply furrowed face of a very old man. There was a gleam in his eyes. What? Did he actually wink at me?

I walked over to him and bowed formally. He made a motion to me to also bow to a fierce-looking warrior to his right and I bowed even deeper to him, which pleased him – being the Chief. The older one was the medicine man. To him I described the construction of a rain meter by making a drawing with a stick in the sand. His perception was amazing. After just a few rough sketches he picked out two men, poured out to them some staccato sounds. Off they went and soon came back with a very acceptable rain meter.

Before the meter was handed to me, the wrinkled medicine man showed it to the chief and the chief's wife and the other elders of the tribe. He explained its functioning by eloquent hand signals, which I carefully memorized.

Then he handed the meter ceremonially to me. I repeated the

description in my own terms, in English, with a few Malay words for decoration – and made the same signs with my hands that the medicine man had made. They understood. The words didn't matter. They projected the medicine man's words, for the signs were the same. The sour and stern chief even nodded and looked a wee bit less grim.

I felt I had become part of their environment and they of mine. I wasn't sure that they did not already consider me an assistant to the great Rain God. The medicine man at least.

I was invited to eat with the Dayaks in their community house. Fired bananas, steamed fern leaves, roasted baby boars and "clapper" milk, coconut to you. Refuse was spilled down through a hole in the floor, every bit snapped up by growling and whining animals waiting eagerly down below. The best disposal system. Cycling. After the meal a statuesque young girl detached herself from her parents, came up to me and looked trustfully into my hungry eyes. She was charmingly naked. I did not know the etiquette, nor what any act of mine would mean to her, to her parents and to the Dayak tribe. I stroked her radiant hair and walked her back to her parents.

Time for bed. It was woven of palm leaves and stretched out between four poles. There was a roof above, but no walls. The "Pesang Grahan" was a hundred feet from the community house, on the edge of the jungle. I fell asleep at once.

I do not know how long I had slept when there was an urge to get up. But I continued to sleep. A battle royal developed between two factions of myself. These two factions began to tear and rip and finally they came apart. One of the parts, the most alive and alert one, who had wanted the whole of me to get up at once, soared up to just under the ceiling and looked urgently and angrily down at the other part still lying on the bed.

The reason that the "I" under the ceiling wanted the whole man to get up was that he saw or knew that the Dayak Chief was on his way to the Pesang Grahan with a drawn hooked sword. And, after all, he also knew that a blond head was worth nine times the price of a black head. There was no fear, only a reasonable amount of urgency. However, there was no doubt that the lazy lump on the bed eventually would get up.

The "I" on the bed also knew of the Dayak Chief's approach. But he thought he could wait just a wee bit longer. He felt pretty tired and he rested well on that woven bed.

With a bound the man on the bed sat up and then stood up and the man under the ceiling quickly and smoothly joined him so that they were one person again – just as the Dayak Chief made his appearance with his sword lifted.

I went smilingly forward with my gun in my hand and bowed formally, asking the Chief in the best English in what way I could serve him. The Chief's sword came down at his side while he bowed in return. He spoke and though I did not understand a word of his Dayak tongue it seemed perfectly clear to me that he said he had come to protect me against any possible attacker, wild animal or human.

I replied in my best English that, while I was grateful, such protection was not required since my gun was so made that it went off and hit any intruder who approached while I slept. He seemed to understand me for he looked surprised at the gun then turned and left.

While I sat on the "bed" a little uncertain about what to do next, the medicine man came along, very quietly, and motioned me to climb the ladder to the community house. There he found me a cubicle to sleep in and he placed his own bedroll across the opening to my cubicle and went to sleep there, protecting me. Obviously, the old wizard looked upon me as an assistant rain god, or better. The Chief's opinion of me was hardly that high.

This night's experience of becoming two from one was never repeated, nor ever attempted or sought, since I saw no need. But there was a follow-up. Thirty years later I began a twelve-year stretch, with interruptions, as a research engineer and finally as a consultant at the University of California's campuses in Berkeley and Los Angeles. This octopus university also had a campus at Davis, where Dr. Charles Tart conducted a study of out–of–body experiences. Somehow he got wind of my long-since feat in Borneo and sent me forty-two printed questions he wanted answered such as – when part of me went out of the body, could I see this part? Were there limbs, a body, and had it clothes on? Did I make a general practice of going out of my body (at least with a part of me) or did it only happen on special occasions?

The questions were clever, many of them designed to trap a liar and the term "out of body experience" was acceptably accurate. If Dr. Tart had called it Astral Travel or Astral projection I would have thrown the stuff in the wastepaper basket. The word "Astral" (meaning connected with the stars) has become a shroud for superstition, deceit and greed, one of these terms that cheapens our language and its users.

When I had wiped cobwebs off my face, I managed to see why a scientist could, and should, ask such a question. I could also see his eager stunt men try never to disappoint the noble scientists and the hungry public. And I seemed to know that the more violent the efforts, the less chance of success. In my own case there were more important factors: My dividing up my personality had been done for a purpose, not as a show, not even as a scientific proof. Whether the act had been done by myself or someone else, I did not know, nor did I particularly care, but to me it seemed a sacrilege, a break of trust, to experiment with this. Someone will say it is ingrained feelings established by family or environment. Hardly. Neither my family nor my environment had trained me in personality-dividings; probably did not even know such things were possible. Rather, my feelings during this for-me unique trip was deep respect and reverence for whomever or whatever caused it and not something to play with for fun or fame or money. No, not even for "science" as we know this term today.

Dr. Charles Tart had great success with other out-of-the-body guinea pigs who felt differently and perhaps better than myself. In other parts of the world this feat has been accomplished quite often, so often that we are required to understand that this is a valid talent or function of the mind, any mind or a few minds – we do not yet know.

This part that separates itself and moves away, is it really mind? Or the part that remains, is that mind? It may not be too important what we call this thing or things, though at the time of separation, what happens does not fit neatly into our conventional concept of mental activity.

Yogis and Sufis, some of whom "leave their bodies" in a manner and with an ease that could be termed professional, use such terms as the Buddhic principle, or the genie sphere, to describe the cluster of energy bundles in and through which separation takes place and they have other units or systems of energy bundles beyond these, which use mind, but are not considered part of the mind.

When these professionals separate their beings, or go out of their bodies, it is usually not for such trivial purposes as urging the remaining part to get up. They may visit friends or disciples and they may do this in subtle, invisible form or even visibly.

What chance has "the man in the street" in this game? Many apparent men-in-the-street do it all the time. It is a talent that some people have and others acquire because they want to, and we realize, sometimes in a sudden flash, that denying the possibility is just as unreasonable as imagining you can do it before you can. There is no need or profit in "believing" before achieving.

An Indian businessman had to travel to London once and sometimes twice a year and was concerned about the expense and the long time he had to be away from his business, for this was before the time of airplanes. He heard about a yogi who was said to be able to transfer himself in no time to any place on earth. How convenient, he thought, and how cheap, to go to London in this manner, to save the long weeks on board ship – the tedious preparations, the passport, the shots.

He looked up the yogi and told him about his idea. The yogi nodded, gave him some exercises, then looked the man over carefully.

"Now go home," he said, "Do the exercises I have given you and don't think about a monkey all the time."

"Of course not," frowned the businessman, offended at this silly admonition. But when he came home, the monkey could not get out of his thoughts. The monkey danced around in his head and almost drove him crazy.

He rushed back to the yogi, "Get this monkey out of my mind!," he bellowed.

The yogi had seen that this man did not have the clarity or concentration of mind required for out-of-body experiences. All he could hope to do with him was to frighten and upset him with this monkey so that he would see the need to train his mind at least to the extent required to get rid of the monkey, if not travel to London.

However, mothers have been known to mind-travel thousands of miles to find their sons, wounded, on battlefields, without any rigorous mind-training; directed only by their deep devotion. And lovers have found each other across oceans.

These energy bundles, this "principle" that can move at least part of us out of our physical bodies is our very closest environment – or is this ourselves? Whatever we call it, it reminds us of and testifies to the depth and flavor of life.

How do the yogis and other "professionals" view their feats? They view these not as a trip, really, nor as a "separation". They do not identify themselves with their bodies but see the whole environment, the whole planet as themselves.

A woman with beautiful eyes may for a time identify herself with these eyes. Then she gets a stomach ache and identifies with her stomach. Later, again, she may become aware of her whole body. So, say the yogis, a man thinking a bit superficially identifies with his little body but when you become more mature and knowledgeable you know that is just a small part of you.

The planet is your body; perhaps the whole universe. So you can transfer your consciousness and even the visible impact of it to anywhere in your planet-body and, if you are real sharp, even to another planet or another solar system or another galaxy.

The mother who visits her son on the battlefield does not, then, "go out of her body." Her body includes that battlefield. She is just transferring her attention to another part of herself.

A shocking aspect of this view: another man is yourself. That is why you can see and experience his thoughts if you really want to, in the proper manner.

As said Mowgli, Rudyard Kipling's jungle boy, "We are of one blood, you and I." And then all the animals in the jungle, to whom he said that, became himself and he them, and they lived in harmony.

INTER-ENVIRONMENT

A scene from World War II will portray inter-environment, how one phase of the environment influences and merges with other phases and vice-versa. So that the sober observer, after having taken the environment apart and scrutinized every bit, puts the parts together again and sees and feels the whole.

A dull winter morning with heavy overcast on an airfield outside Brussels my commanding officer and I, after breakfast in the mess, looked out at a flock of scarecrows. There were about two hundred of them, their ragged, dirty clothes barely covering their emaciated bodies. They all looked sullenly down into the dirty snow, from time to time wiping their running noses and drying it off on their pants.

"What do you think?" mused the CO.

"Who are they?"

"The mayor sent them over. They want to work for us."

Behind or beyond this crowd that hardly seemed to be alive I saw a nation that had been maltreated, tortured, for four long, ugly years and who now came to offer help to their liberators. I went out and talked to them, in my poor and halting Flemish. First they laughed at my funny accent, my clumsy sentences. I told them how grateful we were, also that I was sure they counted among them men who could do much better and faster than our servicemen what needed to be done. At that time I hardly believed this, but they looked up and I thought I could see a glow in their tired faces. I added that, since I did not know any of them, there was no way for me to organize a hierarchy of supervisors and supervised. They would have to organize themselves for each specific task – the men who knew this task coming forward to lead, then dropping back into the class of general helpers at the following task, where new experts would take over.

Their eyes shone when I said this. A bond had been established between us. I had been wanting to see and operate such a unit after watching for years the humbug of the hierarchies at the working places. And my Belgian workers – like workers everywhere – had the same idea. An old man sank his dancing eyes into mine. They were misty grey like the sea fog covering miles of Belgian lowlands. There was laughter in them but no sarcasm. There was compassion with my poor diction, admiration for my courage, wonder over who I was and why I tried so hard.

I voiced the usual warning: "Please, before you join us, think over the risk. Discuss it with your family. You will be in a shooting war."

The old man leaped a few steps forward, "Don't worry, Captain, Sir. The Germans always aim too high."

Another joined in. "Petit is right. To the Germans we little people are too low even to be aimed at."

So I knew his name, Petit, "the little one." French, not Flemish. Other names filtered through as work progressed. For every separate job they formed, as it were, a specific organization for that job, only guided by experts among them in that field. When a new task loomed, these experts sank back into the class of general helpers while new specialists moved forward and up. No one ever asked for, or was given, any permanent rank or title. Petit himself seemed always to be among the general helpers, taking on the simplest tasks while smoothing and lubricating the entire "organization".

They built barracks for our pilots to protect them from colds in the raw Belgian winters so they wouldn't choke in their oxygen masks. German barracks on conquered camps and some even from far inside Germany were torn down, then moved to the proper place on our operating field. Nails were straightened and rust removed before erecting the barracks again. The nails became a bottleneck. Time was at a price. Petit knew of a Belgian nail factory nearby.

We went there and found it in charge of a stern British Army Colonel who bit me off with, "You know I can't give you those nails without a warrant in triplicate from the War Office. What would you say if I asked you for two spitfires?"

I thought we had been roundly defeated as we ambled back to the air field. Not so Petit, "Do you recall, Captain, " he mused, "that we have two irrepairable spitfires in our hangar?"

The C.O. bawled us out when we approached him. Petit discretely caught me by the arm. I thought he meant: Keep quiet. We said nothing to the C.O., just looked humbly at him. After minutes of staring, he gave in.

We loaded the wreckage on tank retrievers and proceeded to the nail factory. I went in and asked the Colonel to come out and inspect the hardware. He looked darkly at me, waddled out and surveyed the scenery. Then he turned to me, "Take your nails – and your spitfires too."

My own contribution to the work was laughably insignificant but whenever a task neared completion Petit would see to it that I was present at the christening and, when possible, make it appear that I was even making some important decision. A few weeks before Christmas 1944 the crew faced a tough job. We had taken over an airfield the Germans had just held. Our infantry had not given them much time. The barracks were sieves from machine gun fire.

Nevertheless, the Germans had utilized their last moments while pounded by fire to utterly destroy a combined heating and water supply plant. The piping and the delicate machinery had been blown up by bombs and grenades. Twisted and torn pipes glared at us from a dark hole filled to the brim with black hulks of anthracite. I expressed my doubts that anyone could clear up this mess. Petit made the simple reply, "Warm baths are good for soldiers."

And so they went to work. No experts were needed at first, just sheer toil. Everyone pitched in until all the anthracite was neatly stacked in covered bins. Then plumbers and machinists took charge with carpenters and masons as helpers.

On Christmas Eve at five in the afternoon, Petit called me, "We aren't sure it will work," he said as we looked over the gleaming pipes in what so recently had been a nightmare. It was still terrifying. All plans had been destroyed. What was I looking at? A haunting dream without substance? Perhaps even a death trap? Walking as in a trance, I fixed my gaze upon one gigantic valve in the center of the maze. Petit seemed to draw me toward it. I placed my hands on it. Before I turned it, I took a last glance around as if to survey the network with the sure mind of a super plumber. Then I turned it.

A hissing sound ensued as of a thousand snakes hurrying toward their victims through jumbles of dry leaves while little featherlets of steam oozed out of hastily connected joints…… "It is working!"

I had to hurry back to the Christmas dinner held on Christmas Eve for the sake of the Polish and Norwegian contingents. There was turkey, cranberries, ice cream and speeches. Everyone was kidded and hounded into participating. When my turn came they shouted, "Captain, King of the Belgians, where have you been hiding? What do you have to say for yourself?"

I rose, slowly, reluctantly, as one who thinks he has better things to do than give after-dinner speeches. I looked up and down the row of Gentlemen-officers whose faces were beginning to gleam with impatience… "Gentlemen, your hot bath is waiting."

There was a moment's stunned silence; then, a roar. Everybody rushed out abandoning further speeches, ice cream and turkey to savour the supreme luxury of war: a hot bath.

There was a prize for the best speech: One month's extra chocolate ration. Mine, though short, won. By dickering and mortgaging future rations, I salvaged enough so every one of the Belgians could have a bite.

The Belgians had one meal a day as part of their wages, the same meal the airmen ate, distributed at their working place. One day Petit told in his beautiful and diplomatic manner, that they were very grateful for this one meal but had been thinking it must mean a while lot of extra work for our cooks. So why would we not rather give them the raw material and they would be glad to prepare their own meals?

The proposal, thus presented, was enthusiastically accepted by our cooks. I was invited to eat with the Belgians at their first meal. As I ate I became more and more amazed.

Finally I winked at Petit, "My friend, you have cheated. This culinary miracle certainly did not come from the humble base of our Air Force offerings."

"Captain," he replied with the straightest face, "This is exactly the same meal you have been eating every day in your Air Force mess. Not an iota added or subtracted. Don't you recognize it?"

I ate with the Belgians every day after that.

One fair day a British military engineering unit arrived. The Colonel in charge told our C.O. he had heard there were backlogs of engineering tasks at our wing. The C.O. pointed out strips and barracks to be built and asked how long it would take. The Colonel conferred with his junior officers, sergeants, and technical experts. He finally came up with, "Oh, about fourteen days."

"Can't wait for that!" snapped our C.O., then turned to me, "How long would it take your Belgians?"

Petit held up four fingers to me. "Four days," I blurted out. Too late! The look on the Colonel's face convinced me I was no longer a gentleman. We did it in three and a half. I am ready to concede that we had all the advantage: Familiarity with the requirements of the unit, with local conditions and, above all, a far-superior organization.

On the day of completion Petit asked to borrow my jeep. Borrow my jeep! After all that Petit and his men had done for us, how could I explain that there is one thing an airman cannot let go of to anyone and that is his vehicle?

Petit was ahead of me again. Realizing my embarrassment he chuckled while explaining that no one was going anywhere in the jeep. It was only a matter of keeping it right outside the hangar for a few moments.

As I left the jeep it was immediately surrounded by a wall of Belgians. Out from the hangar came other workers carrying strange, angular items with which they disappeared behind the wall of fellow-workers. What were they doing to my jeep?

It took only minutes before the wall of workers withdrew. There, instead of my lowly jeep, stood a magnificent two-door sedan with a body of aluminum doors and plastic windows such as no man or officer had ever beheld. "We thought our Captain should not ride around in the freezing Belgian winter in an open jeep," explained Petit.

Just as we seemed to be on the peak of popularity ugly rumors shook our mountain viciously. There had been security leaks from our airfield. The culprit or culprits would obviously be among these Belgian workers whom nobody knew. I was ordered to find out. With the heaviest heart I approached the group. Before I had said a word, Petit looked at me – seriously, this time – not with the usual smile. "Captain, would you please tell me, have you had any security leaks here lately?"

I told him the story, adding that the Belgians were under suspicion. He smiled again, shaking his head, "We know each other well enough," he said, "But look."

I looked where he pointed. A nun in full regalia was just crossing the field. "Yes," I said, "A holy nun," mentally crossing myself....

"No, not a holy nun," said Petit. "A nun never walks alone. Don't be afraid to question 'her.' Do, please!" The 'nun' was questioned. He was the leak!

When New Year approached, Petit showed signs of nervousness. I wondered if he had driven himself too hard. I asked him if there was anything I could do.

"The Germans know every well about your devotion to New Year festivities," he said – probably his diplomatic way of referring to the dead-drunk Britishers he had seen on other New Years' nights. "I don't say the Germans will attack. But I think you should be prepared."

I told the responsible officers of his fears. There was little response except that a few sand bags were stacked around important installations. On New Years Eve Petit and his men puttered around the hangar busying themselves with further improvements. At five in the morning on New Year's day I again saw them at the hangar. I went out. A moment later came the Stukas. They roared around the hangars sputtering machine-gun bullets. Petit laughed up at them. Suddenly he fell in against me and sighed, happily, I thought.

The Germans hadn't aimed too high this time. One of their bullets had found and penetrated the noblest heart.

To the Belgians, Englishmen, Americans, Polish and Norwegians who took part in this trip into a non-hierarchial working environment, this became not merely a quaint war experience but a pointer toward what was to come. The fact that a mayor of a Belgian town had had an idea plus our feeling for our Belgian war victims and their feeling for us – all these things set off this venture which taught us a few things.

I don't think that all the big and weighty organizations now representing hierarchial bossism will suddenly bow to our concept and retreat. No, they will continue on their way faithfully following their long-established principles. The new way, already understood and practiced by many of our young people, will form their own groups and work independently. In the vast ocean of new ventures, new activities that will inundate us when we realize our actual potentials, there is room for all types. The bossists will continue with their weighty instructions into "management principles" and we others will develop and work under exactly opposite management principles. That is the beauty of the United States: We are not bound to or enslaved by one principle, one prophet, one way of doing things; one stupidity. We may have and try all stupidities. Whatever we try, some will call our attempt a stupidity so we may just as well start out with that term.

ARE YOU EMOTIONAL?

This imbecile question, hurled at fellow-men-and-women, sometimes even in psychiatrists' offices, presumes that there are two separate divisions in your mind: Thoughts, and emotions. Thoughts are the sober stuff of which most men think they are made, and all scientists. Emotions are what the women are made of, the unstable, the immature – in some men's views.

When Isaac Newton thought he had discovered gravity but wasn't quite sure yet, and wanted to back it up with some math, he was so uptight he couldn't handle his equations and asked a friend to do them for him.

This friend was no less gripped by emotions, both for himself and on behalf of friend Isaac, but he still managed to do those equations.

Few things have so confused our social sciences and their assumed behavior patterns and our relations to our entire environment as this superstition about emotions. Emotions are the stuff all thoughts are made of and from. The scientist in his laboratory is as flooded with emotions as the lovers in the bridal bed. Behind any and every thought is an emotion, the driving force, without which the thought could not exist.

Emotions are the oceans of the mind. Thoughts are the waves on the surface. Some manage to still the waves, to live in the depth of the mind ocean, void of concepts, void of forms. This exercise is refreshing and may turn you into an originator, also called a genius.

When that aging meteorology professor at the University of California asked this almost equally aging author to apologize, emotions were seething in both of us, yet, we thought we were scientists and pure thinkers. We were scientists and thinkers but not pure thinkers. I may have met one single pure thinker throughout my many years spent in sixty-five countries. I am not quite sure even of that.

When we vote for a President or even for a senator, some hope they are voting for a pure thinker. Others, more soberly, vote simply for the less pure against the lesser pure. Some vote for narrow thoughts and narrower emotions matching their own. A viable government may emerge from all that, for what builds, rebuilds, cleans and runs cities, towns and nations are emotions. And if these various types of emotions elected to office can cooperate in spite of their differences, the work can be done. If they do not insist on excluding colleagues with a wider or narrower emotionalism than their own, they can go somewhere together. If they cannot form a team, the work won't be done.

One is tempted to match Buckminster Fuller with a certain personnel director in our esteemed Department of State who thus defined one of his functions, "Deciding, from the applicant's record whether he (she) is soundly patriotic or harbours vague sentiments about internationalism and one-worldism."

How would Bucky rate there? He never told us but he has a little piece about the English and the Australians in his *Ideas and Integrities*: "Englishmen used to be content to leave it to the Anzacs "down under" to work out for themselves their inherent handicap of having to negotiate all their lives upside-down. We Americans have adopted the same attitude to the Chinese, and vice versa."

The English, as we know, used to send their law-breakers "down under" to be subjected to that awkward routine of living upside-down. These lawbreakers handled the matter so well that they have attracted lawful Britishers and Americans.

Meanwhile, both the British and the Australians have done more for their local ecologies than almost any other nation, and one suspects that their famous understatements come from an overflow of emotions rather than lack of them. Leo Rosten writes in the *World Magazine*:

"Where Egypt wails, England blinks. Where Hindu women tear

their hair, English women study their nails. An Italian explodes invective, an Englishman sniffs, "Really?" When the Russians thunder, "Avery one knows…." the English demur, "But I should think that…." Where Americans cry "It's terrific!" Englishmen concede, "Rather impressive." And where Englishmen murmur, "What a pity," Jews cry, "What a disaster!"

Emotions dominate our family life, social structure, sciences, politics, national goals. These emotions are promptly absorbed by our immigrants from all corners of the world, showing what a coherent nation we are. The trend is pushed by legal and political steps.

An example is the concept of the melting pot, that all the immigrants, of all races, must become thoroughly blended and equalized, as so vividly portrayed and vehemently contradicted by Mayor Kollek of Jerusalem in an earlier chapter. He concedes that each ethnic group should have equal opportunities and be awarded equal respect, but many of them do not want to mix, or even to be bused to schools far from their homes or their parents. Some of the schools established for specific minority groups are among the best in the nation and singularly suited for the groups for which they were established. While some people within minority groups are naturally absorbed into other groups, many are not, and are entitled to keep with their own.

Into this complex and delicate field of minorities and races barge our budding social science, its professionals, hangers-on, and critics. The "intelligence level" of the various races is to be "researched". I am a WASP, white, Anglo- Saxon or Scandinavian, and if not "protestant" then looking like one. Even I know that our famous "IQ" does not tell a thing about me, my visions, my contribution to society, so how could it tell anything at all (except phony notions) about non-wasps? The scholars in question realize this and are seeking new tokens of evaluation, yet, they are already seeking funds to seek and talk about racial "characteristics" eons before any such "tests" can be taken seriously.

The most intelligent man I know, who was involved in a complex maze of military intelligence in World War II, was given a choice of what he wanted to become after the war, and the services would finance it, in view of his excellent record. He chose insurance. They took his IQ. It was 35. They told him with such a low IQ he could hardly make it in insurance. He sold two hundred thousand dollars worth of insurance in one day after having insisted on trying. Today he is a top executive in one of the largest national companies.

An IQ, of course, shows nothing but conformity, willingness to play along with a white-anglosaxon game. My friend was a Cuban. Besides, his wife had just crashed, and died. She was ferrying flying fortresses to the front. Why should he play along with a silly game? To think that professors, teachers, and guides for the young, have such fantastically comic concepts of their "science". Who'd blame the young for rebelling?

One of the tasks of this Cuban during World War II was pretending to be a French colonel collaborating with the Germans. He followed German Military Headquarters all through France and into Germany, taking part in their secret councils and informing the Allies. To keep such a role and live, one has to have eyes and ears on all fingertips and in the back; to feel and know every second the minds and moods of everybody around him; to be, indeed, a superman.

The discovery or feel of danger sets in along the strangest paths. "One afternoon," René told me, "As I was sipping brandy with my dear German friends, I suddenly was beset by an impulse to kill myself. I whipped out my gun, wheeled around, and shot a man poised to kill me."

"Science" rejected this man as an insurance potential. He is leading the pack today.

How did that sudden feeling come to him? So timely? Even though imperfectly, or, in a sense, symbolical?

Did René read his adversary's thoughts? Everybody does, all the time, though most people ignore these hunches and aren't really awake to notice. The 'gift' requires the sharpest intelligence and constant challenge.

What if some of our spies had this capability and could listen in to an enemy's planning? As touched upon in an earlier chapter, we have for years had an office in the Pentagon, headed by a colonel, sometimes a career officer, oftener a psychiatrist provided with a military rank. Whether psychiatrists or carreerists, they told me during conversations through the years that they looked forward, tamely, to a "breakthrough", so spies could be taught telepathy. How convenient, when all they had to do would be listening in to the adversary's war planning sessions. The colonels were patiently waiting for this breakthrough. Breakthroughs never come to those who wait, only to those who toil and sweat and sometimes bleed in eager pursuit of their goal. Russia has such offices too – and China, there this art is a native treasure of millennia.

The military and the universities are not the breeding grounds of the "mind readers." Telepaths are an elusive class. One could imagine they are haughty, insolent or irreverent – for no government , no university and no money can buy their art. It is not for sale. Universities take statistical notes of average people. These statistics are only moderately encouraging.

A person in whom the complex thought patterns around him come flashing into his mind feels no compulsion to talk. His experience is so overwhelming, so sad, so funny, so complete and yet incomplete, so far beyond any cataloguing that he bows his head in wonder and horror. It is difficult for him to pinpoint a plan or a talk. The pictures he receives are so involved, so colorful, so many-faceted, have been transferred in such a way that he feels like a father confessor bound to silence. He would never use the expression that he reads thoughts. The thoughts of others urge themselves upon him.

No field of endeavor is more fraught with traps and tricks. It isn't even an endeavour, for the more you "endeavor" the less you achieve. This is one of many reasons why demonstrations are rarely successful. If you are asked to "prove" any ability to "read" thought, you are already defeated. Thoughts from others pop into your mind when you don't try. In a sense these thoughts don't pop into "your" mind for the mind isn't yours any more when such things happen. You have left "your" mind and have entered a universal mind pool though you may not know it. Even famous telepathy showmen don't usually know this. They feel that whatever horrible thoughts come popping in from someone in the audience, the "reader" can feel no abhorrence, not even criticism. For at that moment he is that other person; lives and feels as he feels, not as he felt himself, just before this transfer of thought happened.

Apart from the Pentagon colonels spending their time waiting for breakthroughs, we have had the card-reading students at Duke University conduct a series of experiments comparable to a study of violin music by giving fifty random students a violin and bow and letting them fiddle. Even if your ears could stand the noise, there wouldn't be much gained in understanding of violin music, such as has been rendered by violinists of the caliber of Jascha Heifetz.

Thema Moss recently conducted telepathy tests at the University of California with new, remarkable features. First, the senders were fired by pictures that evoked intense emotion. Thelma realized that this would create stronger thought currents, more easily read. She also knew the role emotions play in every thought. Another new feature was that none of the participants collected or related the results. That was done by outsiders, some of whom did not believe in the possibility of thought transference.

The humanist psychologists have, at their conventions and seminars even more relevant experiments in these areas.

Different – and perhaps more basic – information in these matters will soon be available from accomplished yogis and others now working in this country such as Swami Rama at the Menninger Institute – if we investigate not merely a narrow speciality but the whole wide range of their field.

In 1923 I met Inayat Khan, a Hindu musician and Sufi, in my native Oslo, Norway. He wanted me to translate his lecture at the University. I came to arrange it all and because of the long line waiting outside his hotel room they let in another man with me. Knowing this man to be rather talkative, I feared it would be difficult to have everything properly arranged. As we entered, Inayat Khan smiled, "Shall we have silence?"

We sat down on a couch with Inayat Khan between us. Not a word was said. That was all the arrangement we made.

Although I had practiced meditation among yogis and others, I am embarrassed to admit that during that silent quarter of an hour nothing else came through my mind than irritation and a thousand itchy questions about what would happen at the lecture without any preparation.

The summer of 1924 I spent at Inayat's Summer School at Paris. Once I came to him to explain about a lady he had appointed leader of his group in Oslo. He again asked if we should have silence. During this silence something happened. The thoughts I had collected about how to present my complaint grew hazy and an entirely new world, sharp and concise, entered my mind like a storm. First I was proud: what splendid concepts had suddenly appeared in my mind! Then I realized it couldn't be me, not quite, anyway. I had never had these thoughts.

Were they Inayat's thoughts transferred to my mind? Or had I been catapulted into a thought pool?

I don't know. How exciting, how liberating, to be able to say: I don't know. If our science and our pundits would be able to say that more often, how far ahead would we be by that one step?

This Delightful Disobedience

To the builders of aerodromes in Casablanca in Morocco came an organizer fresh from the States. This thirty-year-old systems man had provided uncounted business firms with structural back bone. The plans and the systems seemed at first so clear and consistent – the control from the top so firm – the input from below seemed pleasingly muted. Later many of these firms deteriorated. Some collapsed, others reorganized.

This wizard did not talk much to the people in Casablanca. He knew from experience what to do. Some of the people nevertheless tried to talk to him, tried to tell him what they thought should be done, offered their own ideas. To these people who bothered him with proposals of their own, who interfered with his work, he said, "We'll cut you down to size so you, too, will fit into our organization." That calmed the rebels.

A few months later this whole organization blew up in the kind of scandal that happens when orders flow smoothly from above down and the flow from down up is strangled. These type of organizers, who know nothing of what they systematize, except lifeless structures, brought our nation to near ruin, permitted pollution of waterways and the air, ran us into a lopsided economy on the basis of which organizers fired scientists, engineers and economists who had built and enriched the nation, fired them in the cause of "efficiency" and particularly if they questioned the imposed directions, methods, or lack of directions and method. Those who followed the leaders were retained, "if possible." A shining example of this kind of organization is, or rather was, the Soviet Union, where the price of disobedience was too high, so the organizers ruled unhampered and threw their nation into crisis after crisis.

We have faced crises too, because we have followed more and more the example of the Soviet government, or its past, for now today the Russians are beginning to see the folly of their old ways, and release more power, a little at a time, to the people.

A California friend of mine used to tell me, "try to break at least one law every day." He managed to live out his life without ever being thrown in jail, though he did succeed in stirring up some people who possibly needed to be. When our military forces occupied Japan after World War II, Washington DC chairborne organizers sent out fierce orders prohibiting "fraternizing" and particularly forbidding the military ever to pick up Japanese hitchhikers. So my friend Carroll E. F. Nelson picked up four chuckling Japanese school girls and four brave boys every morning and drove them to school while singing with them Japanese and American songs, thus rebuilding the Japanese faith in the world, in themselves, in a recent enemy: the United States.

Our genial disobedience built up this nation's pride and morale and made us a unique world power. It is this same blessed disobedience which, belatedly, is countering our dangerous habits, tackling our ecology, our very minds, in the face of still obstinate resistance by powerful bureaucrats in private firms, in trade and professional organizations and in government. The destructive trends were seen and vigorously fought all along, not merely by such names as Rachel Carson, Barry Commoner or Ralph Nader. Thousands of unheralded engineers, biologists, medics and men in the streets have protested, risked their reputations, lost their jobs. This vicious thing about jobs, unemployment, has done more than anything else to pollute our environment. The majority of those who now run our governments, professional organizations and major companies are still in their hearts against the complete revision of our technology required to re-establish healthy ecosystems, which would interfere with their own beautiful plans and systems. So they often welcome the powerful weapon of unemployment, by which they can intimidate job-dependent and not too affluent employees.

While most management courses still lean to this trend of thinking, there is a new and coming trend expressed by Peter F. Drucker, teaching management at the New York University. Dr. Drucker writes in *Harvard Business Review,* "Within another ten years we will become far less concerned with management development (that is, adapting the individual to the demands of the organization) and far more with organization development (that is, adapting the company to the needs, aspirations and potentials of individuals.)"

At present most people see such development as detrimental to business and profits, which is the reason for its slow take. Forward-looking business firms have already taken the jump, some of the biggest. When more follow suit and the increased efficiency and profits will be evident, then, and then only will the big change come – a change of our attitude toward the total environment.

At the present time, Peter Drucker's ploy is still just another thesis or theory. Is it possible to go deeper into this matter? The shift is not or should not be just from one theory to another. It ought to be from theory to insight. Insight? Yes, into the nature of man, man in his environment. Every single worker, tradesman, professional, executive, president is a man in his environment and until we see them all as such we are working in the dark, clinging to theories, one or the other.

Presidents or premiers who guided nations through difficult years often had intuitive perceptions of their fellowmen's thoughts and feelings, so they were able to gain the confidence of large groups, and represent them. No statesmen represented all the people. Many represented mainly those with base feelings, revenge, envy, anger, and so threw their nation into violent and often disastrous actions. Still, they did have communication with large segments of the population. They acted on a form of insight rather than theory.

Winston Churchill, with his "Blood, Sweat and Tears" struck a responsive chord in millions of his countrymen and represented them truly during that long, tough war. After the war he, and the pretender for the premiership, Clement Atlee, walked in a parade and the throng roared. Churchill thought it was the usual ringing approval of his gallant leadership and held up his hand in a "V" sign. Loud voices in the crowd protested: "We want Atlee!"

They had fought the war with Churchill. Now that war was won and over. They did not trust Churchill to understand what they wanted in a post-war world. They wanted Atlee for that. It was a telling demonstration of men's and women's thoughts and feelings, expressed in an environment where they have a choice, at least once every four years. It was also a demonstration of how a leader can fool himself into believing that once a leader always a leader; or, a leader for one type of crisis being fit also for another type of crisis. These Britishers showed honest and delightful disobedience to a great leader. They discarded loyalty for truth — truth as they saw it.

Loyalty was praised through the ages as a great virtue. It is still praised. Loyalty is what makes an engineer submit to "company policy" and continue to build polluting engines decades after pollution-free ones have been made available. Eventually the company and the whole industry may collapse because of this engineer's loyalty. Loyalty is what makes a child take drugs — loyalty to his group who say, "Everybody does it. Are you chicken?" Loyalty is what makes the young boy drink or smoke in spite of the awful taste in the beginning. He carries on, loyal to his pals — or it may even be to a parent — until the poison has captivated him and made him a slave. Loyalty is what makes a Communist torture a village chief slowly to death even though it makes himself sick. Loyalty is what makes you stick to the old ways even if it stifles your community and your nation. Loyalty is what causes rising nations to stagnate and decline giving room for other nations that, in their turn, stagnate and decline because of loyalty.

Delightful disobedience is what breaks the trend and may cause a declining nation to pick up and rise again.

Loyalty in the area of science and engineering has repeatedly jeopardized our future. May we return to the Ocean Thermal Difference Energy System as an example? In 1956, when the University of California had tested this system for six years and was ready for large-scale plants, first for conversion of ocean water to fresh for the California water supply and later for plants producing electrical energy — another water plan was launched: the Feather River plan. This called for transfer by conventional means of water from Northern California rivers and lakes over hill and dale to the thirsty South. The cost estimate was one and one half billion dollars, an estimate that was later to be overrun. The desalination plant was conservatively estimated to cost five million dollars, and would have produced five million gallons of water a day – just a beginning, of course, but by building this plant, the system would be proven, and we could have gone on tapping the free energy available in the ocean temperature differences to any extent desired.

A California legislator-banker who heard about this phoned me and fumed, "Why do you engineers propose a one and a half billion boondoggle when we can get it all free from the ocean?"

A University of California faculty member from the Los Angeles campus had unknowingly answered that question earlier. Even though he well knew of the planned conversion plant, he pleaded for the Feather River Project with these words, "There is such an overwhelming sentiment for the Feather River Project among the State's engineers, I think we have to go long with them."

There was "an overwhelming sentiment," yes, for choosing a project that the overhelming number of engineers knew and understood. If the other project had been chosen, even if millions of dollars would have been saved and even, in addition, if a pathway would have been opened to a solution of our entire energy problem, many engineers would have been faced with a technology they did not master. So what might happen to their job? Personal insecurity stared them in the face. So how could they be blamed?

Can any one be blamed? Can we blame that Los Angeles Campus faculty member who thought obedience to his erring but numerous engineering colleagues was more important than his own professional conviction? Can the citizen be blamed, that "man in the street" who is supposed to govern this nation and himself? Should he keep informed so he can choose and side with the right expert, not simply with any expert, or even with a numerical majority of experts? Particularly, should this "man in the street" study those economics experts who claim we can have and must have guaranteed full employment, so an engineer does not face unemployment and indignity if he does not master a new and upcoming technology?

In all fields we have a few excellent experts who can get us out of any jam we happen to be in. But not all experts are excellent, perhaps not even the majority. We did not educate experts to take the choices away from us. We educated them to carry out our own choices, after we have made them.

May we return to Barry Commoner and his population explosion opponents? Through a fortunate combination of intuition and science Barry reached essential conclusions about the post World War II technology as the chief culprit in threatening our ecology and how to revert the trend and repair the damage.

Into this discussion crashed the population-control movement which had already become a powerful lobby of a few scientists and a lot of salesmen. They rebuked Barry publicly for not having included the population matter in his treatise. They approached him personally and asked him to change his story, not only to include the population growth as another cause for pollution but to soft-pedal all other causes so as to achieve the "right impact" on the "public". There is something logical and in a sense admirable in all this. The loyalty to the population cause is seen as saving humanity. Even if a little lying is added.

Barry Commoner (and I for that matter) have access to the same books and the same reasoning that the population people have access to. I don't know exactly how Barry feels about all this but my feeling is: I don't know whether a limitation of the population growth will have a total good or bad effect. One bad effect will be that with fewer people around there is less chance of a sufficient number of bright ones that would see the vast resources available and help realize their utilization. The very fact of more people to feed and house would force the solution. Earlier. Particularly is this true of a nation such as the United States. In many other nations the population increase might be slowed down with advantage. That is up to these other nations.

Barry did not fall for the plea of changing his thesis. He saw that the more profitable way now was to so revise and rebuild our technology that pollution could be properly controlled. The proposed "loyalty" to the population scientists and the population propagandists might dilute the matter and delay the necessary action. A scientist, like everybody else, must reject loyalty and work for truth as he sees it. Has any scientist a right to conceal findings from the public out of a feeling of loyalty for his peers? One would rather say that a scientist, particularly, has to practice disobedience, all the time.

Simultaneously with the population scientists' attack on Barry Commoner, the worldwide movement for restricting populations began to dissolve into regional differences. In some countries a decrease came by itself, such as in the United States. Other countries have empty spaces to fill and economic goals to reach that demand increased population growth, as seen by visitors and United Nations missions to Brazil, Bolivia, Kenya, Tunisia and other countries.

The ease with which the case of population growth versus food production has flip-flopped during the past ten years ought to have warned both scientists and the general public about the need for more patient research and spadework before dramatic statements are being pronounced. In the area of energy the judgment must be more severe. Here we have already done patient research and spadework but it is ignored. In the same year that hundreds of the nation's top technical experts pronounced the Ocean Thermal Difference Energy System preferable to all others, particularly in the United States, and ready to build now, today, not merely several politicians but two of the most prestigious scientific organizations talked about solar energy (in which they included the ocean energy system) as just a dream or maybe something to count on in the next century. One begins to wonder what sort of motives enliven so many "respected" scientists and other citizens. Is it again just a matter of misplaced loyalty to a peer group? A loyalty that may yet destroy us as a nation, and possibly most of humanity in the bargain?

The term "disobedience" means here refusing to obey when occasion warrants. It does not mean trying to make others obey you, particularly not by acts of violence or force. Only the weak and confused strike out, bomb, burn, frustrated by lack of skill in achieving.

Demonstrators are in another league, though they also are fooled and hoodwinked and made to shout and express credos most of them never intended to express. They make an impression upon spectators they never intended to make. It is a naughty confidence game, contributing nothing of enlightenment or progress. Mass demonstrations are not the kind of disobedience that is delightful. Permissible, sure; profitable, no. The effectively disobedient person usually stands alone. Yet, he may change a nation, the world. Mass demonstrations can't. Their effect is wobbly, confused.

Comprehensive Designer

Comprehensive Designer, or "Generalist" are terms of our language, owned by nobody. Yet, most of us would relate them to Buckminster Fuller, who may have been the first to use them purposefully.

This chapter will in part be a eulogy to Bucky, who is exactly my age, who has been my lifelong friend though he may not know it, whose books have adorned my shelves and who said of pollutants, "They are unharvested wealth." (*Life Magazine*). Any better definition, anyone?

And what is wealth? In Bucky's words: Wealth has two constituents – Energy and know-how. Energy is constant (in the wider environment), cannot increase or decrease. Know-how can only increase, does so every time it is used or applied. So wealth, too, can only increase (*Utopia or Oblivion*). How, then, could humanity possibly be destroyed? Not by any shortage in nature, but by clinging to old concepts developed under earlier conditions; by superstition, by short-sightedness. All that is needed, then, is an education that will prepare man to live and act in the world we have now. Soft-hearted Bucky, however, gracefully tells us he does not propose to reform man, but rather to prepare and build a technology that will suit him and to which he can adjust.

On education Bucky presents an aside that must be faced here. Referring to Dr. Benjamin Bloom of Chicago University (*Utopia or Oblivion*) he claims that a person's IQ can be determined with one per cent accuracy from birth to seventeen years of age by simple information about the person's environment, parental status and so on. Have we the Marx-Lysenko environment theory here as op-posed to the heredity theory or my own lifelong experience that neither environment nor heredity count for much of a person's ac-tions or accomplishments in life, particularly if that person is strong and talented?

The explanation may lie in the term IQ. For what is an IQ? Sim-ply a measurement of a person's will to cooperate at the time the test is taken and even then it pictures just mechanical, non-essential traits or talents – none of the deeper, creative forces in the person. This becomes still more pertinent when Bucky goes on to describe how "at the age of four" fifty percent of a person's capacity to improve his IQ capabilities has either been expended or protected. Between eight and thirteen years twelve percent more of potential capability is actuated and at seventeen one hundred per cent.

If anyone would imply that man does not increase his talents and abilities after seventeen he is sorely wrong. As for myself, I have learned faster and better and increased my store of capabilities more after 75 than at any other time period.

Exactly the same may be said about my contemporary, Buckmin-ster Fuller, who stated at a later occasion, "Every time man uses the second constituent of wealth – his know-how – his intellectual resources automatically increase."

The admirable author of these words apparently is no longer spellbound by current theories of the IQ and other mind twisters. Awake psychologists like Robert Ornstein of the Langley-Porter Clinic in California see in the ancient yoga and sufi traditions some beginnings and a few definite answers to burning questions our psychiatric science has so far raised but not answered. Buckminster Fuller may have soaked up wisdom and improved his view of man during his extensive work in India.

René, our superintelligent spy of a previous chapter, is today sit-ting high above the Los Angeles traffic, a trusted executive of a leading insurance company. At a recent visit he assured me that if I ever was out of a job he would hire me as a salesman. Obviously a knower of men. A genius with IQ 35.

With the above ambiguities safely removed, Buckminster Fuller's contribution to our future and his guiding light will surely stand out more clearly.

Bucky's labors of love have sprouted many fruits: He found and trained generalists, comprehensive designers. He prepared our society for accepting and employing these generalists. He composed and showed us, in outline at least, his own concept, his own design for the future of Spaceship Earth.

Bucky's generalists were not the first. We have had them all along. Kings, emperors, presidents (of nations, of business firms and colleges), premiers – all were supposed to be generalists though recently they have been surrounding themselves with such formidable staffs of non-generalist specialists that they hardly fill the term any longer.

Cabinet officers were also supposed to be generalists within their rather extensive spheres but they, too, have fallen for specialists, watering down the very concept of generalist. The same applies to an appalling extent to senators and congressmen. All these are now elected or selected not according to their generalist talent but for geographical-political-how-large-contributions to political campaign considerations.

There have been other real genuine generalists; people who covered several – if not all – specialties. For example, Dr. Jacques Ménétrier (eulogized in Chapter 4), a French medic and also an electronic expert, physicist, mathematician, artist and writer who managed to delve into and solve basic problems of his nation such as juvenile delinquency; the tiredness of executives – and, with others in his Centre International de Recherche Biologique: restoration of the French bread and agriculture in general and basic research into the human system.

To operate and improve the running of societies and nations we have economists, but these will not get beyond their clay-footed theories if they do not understand a great deal of engineering and farming. Or vice versa, an engineer or a farmer who also understands economics can perform similarly. And there was England's John Maynard Keynes who had such wide-ranging interest, such strong and true feeling for and with all professions and trades, that he became a real generalist though not academically trained in any other field than economics.

All these whole or half-way generalists, however, have had little effect up to this time, because nations and their leaders did not use their talents, did not understand the need for them, except in a few cases, such as that of John Maynard Keynes. The "specialists" enjoyed almost exclusive respect and veneration, notwithstanding their awkward and sometimes catastrophic dispositions.

What about the arena where events are supposed to be foreshadowed, and future improvements suggested – the magazine-and-publishing world? This world didn't do any foreshadowing. It was as steeped in the idiocyncracy of the specialist as politics, business and most of education. Until recently only specialist books and articles were accepted, were considered "publishable" or profitable. Generalists had as little chance to have their views published as they had running nations.

Buckminster Fuller managed to raise the generalist to recognition and prominence in the face of a prevailing abject surrender to the narrow specialist. In addition he provided a grand, comprehensive design, unburdened by details, of Spaceship Earth's future, overcoming fierce resistance from all national and sectional interests and traditions.

To succeed in these ventures he could not afford the luxury of a "neutral and unbiased attitude", the very essence of current wisdom. In the words of A.K. Talbot, and ardent researcher in a related field, "If the researcher wishes to advance beyond the vicious circle of endless repetitions of half-satisfactory experiments, which are the usual reward of the half-convinced experimenter, he must identify himself and become at one with the situation by adopting, in a way of a working hypothesis as it were, a whole-hearted acceptance of the phenomena at their face value, regardless of how much this deliberate act of acceptance may outrage his intellectual convictions."

Cagely Mr. Talbot adds, "While thus acting as a whole-hearted believer, keep a corner of your mind alert, watchful and unemotional, avoiding all partisanship. A pretty little piece of mental acrobatics!"

While contemplating Buckminster Fuller's comprehensive futuristic design, another effort comes to mind: Julius Stulman's "World Institute", an organization that goes a little more into detail, about housing, food, economics, general attitudes. The two efforts complement each other. Buckminster Fuller appreciates this.

One might term him, therefore, a pioneer who realizes not merely his assets but also his drawbacks from a public and publicity point of view.

This recognition and this attitude help enlighten also those who cannot take Buckminster Fuller literally or directly.

When can all this be accomplished? Sooner than some think because only a modest number of comprehensive designers will be required and a relatively small number of helpers or appreciators. It is important, however, that this modest number be well represented in the national administration, in Congress, among business executives and university administrators.

And of course there will be computers. With the mere mention of computers most people think of large, complex systems; giant machines that can handle all problems covering several states. Not only the general public but many leaders in government and business still think of computers in such terms. However, modern computers rather seek to adjust to smaller, limited individual cases. The latest computer fashion is to be special, simple, accommodating, listening to the user and what he wants rather than, as it was before, the user having to listen to and adjust himself to the computers.

Therefore, the individual approach to ecology, the farming out of specific tasks to private companies and associations, may be served by computer systems with appropriate software all along. The software is the computer face to the operators, the outsiders, and the gadgets that combine and make the results readable and often pleasant.

The most recent developments in computerdom form a mighty tool for adequate pollution control.

This present development toward smaller, specific, almost individual computers conforms with the American dream of free, individual enterprise – a dream that has seemed so often to be violated lately, not the least through huge computer systems seeming to master our destiny rather than serve it.

As we are taking a new look at these individual, adjustable, one might even say "listening" computers, we might as well take a new look at the structure of what we call free enterprise. Too often this is confused with some large companies that have bureaucratic structures not very different from those in huge Government agencies. The meat and substance of free enterprise is something entirely different: a huge network of small companies, many of which contribute enormously to the growth and strength of the nation. At what price? Often such a beginning company makes no profits at all and their executive or executives may live on previous savings, hoping, of course, for increased income later, but often even that does not come. After a company has contributed its priceless assets, it may go broke and its executives and employees find their way to other companies.

I was a proud director of one such company in the forties, that built dynamic balancing machines with electronic registration for the US and Australian navies, Waterbury Clock and others. It was the first device of its kind, built according to French patents. My annual income from the company during these years averaged $2,400, and even this modest sum became available only after we had sued one of our customers, who wouldn't first pay what we thought he had promised. We finally sold the procedure to a larger company covering a wider variety of products and I joined a University as a research engineer, later becoming a consultant, earning as much in a day as I had previously earned in a week in my "Private Profit System."

How much of America's riches and well-being are due to such hardworking men and women forsaking profits for the joy of creating? Perhaps half. Just an uneducated guess. The people manning these daring ventures roundly cuss the Governments and their forms and rules which truly are almost ruinous for these small firms. Many of these people could never work as "civil servants" and would be lost in a socialist country thriving on mock hierarchies.

These ingenious workers and executives in small non-affluent companies are our first, best, and also last "generalists" and "comprehensive designers" of yesterday, today and tomorrow, more numerous and often more capable than those now specifically trained as such. We have all these people here, now, with us. This is why the United States is better prepared to tackle the agonizing problems of our environment now, today.

BIRD'S EYE VIEW

The *World Environment and The World Bank*, a booklet published by that giant octopus with financial arms reaching far into every nation, repeats what Buckminster Fuller, Dan Coughlin and Dick West Brooks have already said, that ecology is profitable business. The World Bank invested eighty million dollars to clean up the Tiete River in Sao Paulo in Brazil. This work will increase the value and use of properties along the river to such an extent that thirty percent return on the investment is anticipated.

Three further examples of World Bank ecology projects now under construction will show what we already have done and can do and at what cost. The cases were presented by Robert MacNamara, World Bank Director, at the Stockholm ecology conference and quoted in the *Bulletin of the Atomic Scientists*, September 1972:

> In its funding of the expansion of a steel plant in Turkey, on the Black Sea, the Bank cooperated with the borrower in building into the specifications – as a result of thorough on-site study – provisions to control within acceptable levels the flow of liquid wastes into the sea, and gaseous effluents into the air. Originally no such controls had been contemplated. The study convinced the borrower that this would result in unacceptable damage to both offshore waters and the surrounding terrain, and the recommended pollution-control technology was adopted. The cost for providing this important protection for the environment, as well as for the health of the local population, was only two percent of the overall project costs.

The second case: in the Yagoua district of Cameroon, the rice farmers are poor. The Bank's estimate was that their cash income could increase fivefold in a decade if only irrigation facilities could be improved. But a serious environmental hazard had to be reckoned with: Bilharzia. This water-borne disease is carried by the bulinas snail, and is endemic to the area. Though the proposed irrigation network would serve 3,000 hectares of land and 2,800 farm families, it was feared that the project might significantly increase the incidence of illness. To assess the problem, the Bank sent a highly qualified expert in the control of the snail vector to Cameroon. After on-site research, his report recommended changes in the engineering design of the canals, provision for periodic surveys of the snail population, and appropriate molluscicide application as required. The borrower welcomed these recommendations, adopted them, and during the loan negotiations further agreed that public health officials would carefully monitor the region. Thus, an urgent development project was protected from potential ecological risk by inexpensive and practical preventative measures.

The third case: in its financing of a marine terminal at Sepetia Bay in Brazil – as part of an iron ore mining project near Belo Horizonte, and its attendant rail transportation to the sea – the Bank commissioned an ecological team to study in depth what was required to keep this unspoiled estuary free from pollution. The Bay supported an important fishing industry, and possessed tourist and recreational potential. The Bank's team included a marine biologist, a shellfish expert and an oceanographer. Their recommendations have been built into the loan agreement, and provide for protection against oil and ore carriers flushing their huge holds in the bay, contingency equipment for accidental oil spills, solid waste handling and terminal sewage treatment facilities and landscaping to preserve the aesthetic values of the area. All of these measures – which will insure that the fishing industry can survive and the bay remain a tourist and recreational attraction – represent less than three percent of the total project cost.

To these examples from the World Bank may be added the case of London, where eighty percent reduction in smoke emission and forty percent reduction in sulphur dioxide have been achieved over the past fifteen years, resulting in a doubling of the average hours of winter sunshine. This improvement is rumoured to have cost each Londoner 40 cents a year.

These examples show that current, basic ecology problems can be handled, and at only a fraction of the cost forewarned in ponderous public cost estimates, that may have been concocted with the specific purpose of scaring the ecology-minded public from persisting in their demands. Also, in these examples the results were obtained by adjustments to the technology without any regard for the population increase, which, specifically in the case of London, went on its merry way all during the improvement. You want to bet on Barry Commoner over his opponents?

Bits and pieces of news from all over the world, picked up at random, may show the trend. We are indebted to innumerable publications among which *World* again stands out, magnificently.

France has instigated design and production of electrically-driven trucks and cars for all cities. Gas-driven cars will not be allowed inside cities. Private companies and the government share financing. In Chile a community of people from various countries and with appropriate skills will be formed as a part of a research project by the Lancaster University. External researchers of various disciplines will monitor the experiment which will try to establish a self-organizing "ecologically-viable" community. The United States has seen many such communities established though none professionally organized, even though psychiatrists in California have investigated a number of such communities after they had been established, especially to see if the children from such units would fit into our school system and later, in society as a whole. The findings were generally favorable.

In Russia, along the Volga, there have been bitter fights between the dam and water supply people and the fishermen. In that equation one may say the fishermen stand for ecology.

Dams in the Volga have been catastrophic for the fisheries. Now

the official papers and powerful factions of government are beginning to side with the fisheries – and the fishes.

In cities all over the world car traffic has been restricted and arranged in fanciful patterns that sometimes gives the impression that
the intention is to discourage the driver from entering the city. In
New York there may soon be a fifteen-block-long mall along Madi-
son Avenue where only pedestrians are admitted. Bremen in Germany and Goteborg in Sweden are divided into quadrants with the
only access from a ring street around the town. This has reduced
traffic substantially (maybe with a few cuss words added from un-
wary drivers) and has also reduced the accident rate. Florence in
Italy has banned driving from a forty-block center area. Altogether
thirty cities in Germany have initiated driving bans in certain areas.
Similar bans exist in Tokyo, Paris, Amsterdam, the Hague, Rome,
Zurich and Copenhagen.

Practically all nations bordering on the North Sea and the North
Atlantic have agreed to curb dumping of industrial wastes in these
oceans.

Among the organizational bodies now concerning themselves
with pollution in a partly-overlapping scramble are: The Organization for Economic Cooperation and Development; the Council for
Mutual Economic Assistance; NATO; the Council of Europe; the
European Community; the UN's Regional Economic Commission
for Europe. All these were already operating when the highly publi-
cized Stockholm Convention got underway.

NATO as well as single nations have been criticized for tackling
the basically international matter of ecology. NATO leaders rightly
replied that such limited-interest organizations are so far the strong-
est among us and can achieve faster results than wider-ranging "neutral" bodies.

The secretaries of these organizations are reported to meet and
communicate and try to coordinate the entire effort. On top of all
that the Dan Coughlins, the Dick West Brookses and, above all,
super-international Buckminster Fuller wield their swords.

With all these organizations and individuals working on the ecology, can we others just lean back and relax?

This line of thought is what defeated so many promising beginnings. All the organizations and the individuals expending their efforts for great causes derive their strength, direction – and funds – from you and me. The minute our enthusiasm cools, something happens to the entire effort. The toiling organizations do not die, do not cease to exist; they simply change into self-perpetuating empires that no longer serve the public or the cause for which they were established. They serve themselves.

For at the heart of all our efforts is that invisible little bundle of energy, excitingly fragrant, like a rose, which we call soul or spirit, or personality. Around it: the environment; first and closest: roaring, battling bundles of energy called mind, including seething emotions that twist and turn the rest of the mind. Can we make our spirit command this story ocean? That is the first hurdle. Success is somewhat dependant on wider environment, parents, playmates, schools, access to meadows, trees and animals. A strong and healthy person may thrive even in a meager and difficult environment.

The next environment, beyond mind, though not containing mind, is a partly visible, bubbling, breathing or wheezing body that has a head and brains, eyes, ears, mouth, nose, neck, torso, tummy, back, arms, legs, all of which have to be cleaned, fed, exercise; influenced by a wider environment, parents, teachers, neighbors, playmates. The body has to have shoes and clothes, or does it have to? Some insist if we ran naked from birth, even in snowstorms and blizzards, our bodies as well as our minds would do better. Gypsy tribes dip their babies in ice cold water for "hardening".

The baby is hardly out of the mother's womb before it begins to breathe, and what does a big-city baby breathe now? The air over cities needs more and faster attention than all working organizations and individuals have offered so far, or have been permitted or authorized to offer. There is recognition, and a few cities, such as London, have acted. However, the way most city babies have to breathe today is a crime comparable to torture and murder.

The way a baby reacts to such foul air and other hazards of the environment is influenced by other environment inputs. With the healthiest first food: from a milk-giving mother, an average baby can stand more abuse than one less lucky.

We grow up, for better or for worse, and see a truck roar down the highway; marvel at its ingenious machinery created by human hands and minds, wonder why such ingeniuous minds permitted toxic gases to bother our breathing. But wait: as early as 1959, at least, we had invented remedies. Why weren't they applied?

Most of us didn't even know about it. The others who knew and could have acted, what kept them? An executive cannot improve his product at a price his shareholders refuse to pay or he would be thrown out on his ear. A pressure must be mounted that changes the view of those shareholders. This is where you and I, the man in the street, come in. We are still the bosses here in America. If we accept our responsibility. Our constitution gives us that privilege. It is a privilege only when we use it.

What do we have now? A jig-saw puzzle in the process of being solved? Or a spider web greedily woven by a ferocious beast of prey, to catch and eat us flies?

Our own minds or souls decide all that. We can solve the jig-saw puzzle. We can control and direct that beast of prey so it behaves and acts as our diligent servant.

This tale began in the staccato noise and under a mushy overcast of suspect fragrance in New York City on the corner of Wall and Broadway. Months later my scene had shifted to a shimmering beach, mile upon lonely mile, between Oceano and Guadalupe on the Pacific, with pelicans and a crystal sky for company. There were shacks built by the moonshiners of the twenties and dug out of the sand by hardy hermits. In one lived Peter, the gardener; in another Elwood, abstract painter, also a meditating yogi, who had a framed sign on the wall, "Remember she who chose the better part". But Elwood had one peeve: Moon Mullins, his neighbor one mile north, "That reprobate, he runs up and down my beach every morning, stealing all my timber."

Moon Mullins was a fugitive from a sheriff somewhere (there were none in Oceano then) and enjoyed his freedom and Pacific rolling rollers except for one agonizing peeve: his neighbor Elwood, a good mile to the south, "That holier-than-thou so-and-so, that abstruse albatross of a smear-painter, shooting up and down my beach each morning, snitching all my driftwood…"

A mile north of Moon Mullins was Moy Mell, a luxury cottage with a shower and bridal bed. It had been built by Chester Alan Arthur III, Gavin to his friends, grandson of President Chester Alan Arthur I. The grandson had been a dunite once and now lived in Hill House, a cluster of expanding structures on a lazy hill overlooking the town of Oceano. He had lent Moy Mell to John Wingate, a tall, bronzed hero of the dunes, of the virile British family of daring noblemen-soldiers. And his woman Emily. They later married.

Most dunites had a lust for life, except Pat, writer and former newspaper man, whom nobody would hire now, because of his flamboyant bitterness that had grown to uncontrollable dimensions just because nobody would hire him. So Pat hung himself in the large eucalyptus tree at Moy Mell. This tree faithfully absorbed and has since faithfully guarded a record of the thoughts and emotions of Pat has he passed through the tunnel – his courage and his fears, his self-pity and his rancor, his bold plans for the future of MAN and his vengeance; his last-second regrets. His mind was a mirror of what all jobless people feel from time to time, now faithfully kept by this large tree, swinging its branches with the free breath of the wide Pacific Ocean.

Last year I looked over this un-owned property again. Both Elwood and Moon had gone. Where they used to live are only pelicans now, turning their eyes cagely as you move along, watching you lest you steal their clams. When you lie down for a sun bath in a cove, a dune buggy roars over the hill, nearly crushing you and letting out a vapor trail of suspect fragrance which, however, cannot yet poison this crystal air.

Yes, there is still more than standing room in America.

The large eucalyptus tree at Moy Mell was still there, brooding over Pat, reflecting his sorrow, mirroring the despair of all those sentenced to one of the worst tortures of the centuries: involuntary unemployment.

Peter was still on that beach. He took care of the beach houses of the millionaires who only come when they want to get away from it all. Peter was frightened now. The Park people had made moves to throw him out.

I spoke to the Park people. First they seemed pleased to chat with a research engineer who was going to take care of the ecology and who had been a dunite in the long-past. Then I opened up on the matter of Peter. He was living a useful life, also perhaps the only kind of life he was able to live. Now the darn-tootin' bureaucrats were bothering him. Did they know anything about this? They cringed. I went after them, asked how they would feel if their own positions and ways of life were threatened, as indeed could well happen, for people were getting restive and might vote nay on the whole civil service system (of which I happened just then to be part) – if this system did not know any better than killing what America has been all about: Freedom to follow your own star and living your own life.

Such freedom is possible only in an environment of consideration, which is now being impressed upon us, echoed from all corners of the globe … "Planet Earth, itself, now demands what the sages and prophets shouted in vain."

Afterword

Is this a book about energy, economics, and the environment or simply one man's way of making sense of the influences, ideas and experiences that have formed his life and world-view? Is there any difference?

An engineer who was a mystic. A scientist with fully developed intuition. An economist and numbers man with an expansive heart. Definitely anti-expert and anti- hierarchy, Beorse embodied innumerable such seeming contradictions, in a life dedicated to obeying his intuition in answer to the call of humanity.

In this last book that he completed during his lifetime, Shamcher Bryn Beorse offers an overview that can be seen from many points of view, each revealing new aspects for the reader to develop, uncover and explore.

Time has passed since the initial writing, and many ideas in the book that had seemed unlikely have become commonplace today. Some of the references may seem dated, yet the concepts behind them are unchanging, so they have all been retained.

This archival edition of *Planet Earth Demands* has been published as it was originally written in the 1970's, including revisions added by Beorse in early 1980. It is one of the publications of the Shamcher Archives, dedicated to preserving and publishing the works of Shamcher Bryn Beorse.

We are happy to note that OTEC is at last finding its way into the wider mainstream discussion and implementation of available solar technologies.

More info at www.planet-earth.shamcher.com

About the Author

Bryn Beorse (Shamcher) (1896-1980) was the author of many non-fiction books, novels and articles, covering topics of energy, economics, full employment, and global awareness as well as yoga and Sufism.

Born in Norway, he worked and travelled in over 65 countries in his lifetime, and he eventually settled in the United States. Fluent in several languages, his comprehensive worldview included the inner meditative life as well as the accomplishment of life in the world. Sent on a UN economic mission to Tunisia in the 1960's, helping to rebuild the Norwegian economy after WWII, Beorse also spent time in exploration, travelling to the Kumbha Mela in India, living as a beach bum in the dunes of Oceano, and going to China at the time of the revolution. A spy in WWII, he was part of the plot to kidnap Hitler. An advocate of the giro-credit economic system, he spoke out against the stagnation of hierarchical organization.

An accomplished yogi and Sufi, Shamcher was instrumental in developing Sufi centres throughout the world, in the tradition of Inayat Khan. He devoted the last years of his life once again to promoting OTEC, Ocean Thermal Energy Conversion, the source of benign solar power from the sea.

www.ingramcontent.com/pod-product-compliance
Lightning Source LLC
Chambersburg PA
CBHW052134270326
41930CB00012B/2878